W9-CPL-427

ASCPL DISCARDED
14 DAYS
Oversize
778.37
Ed36m
A830787
Akron-Summit County
Public Library
Akron, Ohio
PLEASE DO NOT REMOVE
CARDS FROM THIS POCKET
MAY 1 4 1980
FINE ARTS

Moments of Vision

Coronet

A diadem, decorated with pearls above the rim, produced by a drop of milk falling on a plate covered with a thin layer of milk. In the land of splashes, surface tension fashions delicate shapes too ephemeral for any eye but that of the high-speed camera. This print is one of the classic photographs in the history of photography.

The MIT Press
Cambridge, Massachusetts, and London, England

MOMENTS OF VISION

The Stroboscopic Revolution in Photography

Harold E. Edgerton and James R. Killian, Jr.

Pictures in this book taken by photographers other than Edgerton carry a credit identifying the photographer. All pictures undesignated by a credit line were either taken by Edgerton or taken in his laboratory.

This book was set in Linofilm Futura by Wrightson Typographers printed by Halliday Lithograph and bound by Publishers Book Bindery Inc. in the United States of America.

Second printing, September 1979

Library of Congress Cataloging in Publication Data

Edgerton, Harold Eugene, 1903-
Moments of vision.

Bibliography: p.
1. Photography, High-speed. 2. Stroboscope.
3. Edgerton, Harold Eugene, 1903- I. Killian, James Rhyne, 1904- II. Title.
TR593.E33 778.3'7 79-11647
ISBN 0-262-05022-6

Dedication

To all who have helped me in the development and application of electronic flash lighting. Seldom has anyone been as fortunate as I in my friends. The list is long! "Thank you one and all!"

I think that you who have been a part of the action will be happy to see the story and pictures that are assembled in this book.

Harold E. Edgerton

Contents

List of Photographs

The light which we have gained was given us,
not to be ever starting on,
but by it to discover onward things more remote
from our knowledge.

John Milton, *Areopagitica*, 1644

Papa Flash and His Magic Lamp

James R. Killian, Jr.

This book is about Harold E. Edgerton and his achievements in the related fields of stroboscopy and electronic flash photography. It is mainly a portfolio of photographs taken by a technology brought to perfection by Edgerton and used to enrich our vision, to reveal to us phenomena never before visible, and "to discover onward things more remote from our knowledge."

The book, however, is more than just another photographic album, although it contains a gallery of memorable pictures. It is a sketch for a portrait of a world-famous scientist-engineer whose greatest invention is himself. Gjon Mili called him an "American original." A professor of optics at a sister institution to MIT described him as "the kindest, warmest scientist and engineer I have ever had the pleasure to know. . . ." Jacques Cousteau and his staff, with whom Edgerton made ten voyages on the *Calypso*, dubbed him with affection and admiration, "Papa Flash." A distinguished Russian academician wrote of Edgerton's "noble and fruitful career." As his friend and colleague for over thirty years, I sum up my own appraisal by recalling a phrase used by Bernard Berenson in describing an admired friend—"he is a masterpiece."[1] His triumph has been his own life.

It is this prescriptive friendship for Harold Edgerton that led me to say yes when he suggested that we join in publishing another book. I have made absolutely no contribution to Edgerton's photographic technology, but I have found satisfaction in our collegial association as well as in the pleasure of "sudden wonder" in his pictures. But quite apart from my enthusiasm for seeing his photographs gain the widest possible recognition, I eagerly welcome the opportunity to pay a personal tribute to "Doc," as his students and associates affectionately call him. He is one of my MIT contemporaries whose gifts, accomplishments, and personality indeed make him a masterpiece, and I have enjoyed proclaiming my admiration even though the labor of doing so invades the serenities of my retirement. In a way the result is an unconventional Festschrift honoring Edgerton in which the tribute includes an array of the scholar's own work with a commentary by me and others.

It was Edgerton's good fortune to attract another enthusiast for flash photography who greatly widened Edgerton's audience and became a close friend. This was Gjon Mili, a senior electrical engineering student at MIT when Edgerton was a first-year graduate student. Mili first aspired to become a motion-picture photographer, and, after graduating from MIT, went to Westinghouse to join a group working on electric lamps and their uses. There he became involved in testing the then-new chemical flash bulbs and gaseous light tubes and working with professional photographers in New York. The Boston Illumination Society invited Mili to talk about lamps for photography at a meeting scheduled at MIT. Edgerton was on the same program with a demonstration of the electronic flash system. When Edgerton offered to bring a set of strobe lights to New York City for trial, Mili borrowed a studio and arranged for a dancer to pose. This was the beginning of a long association with Mili.

Mili asked, "How can I get a set of these new lamps?" Edgerton replied with a question, "What will you do if I give you a set?" The immediate answer was, "I will quit my job and start the first studio with strobe lights." And so it happened, but with numerous conferences, changes in plans, and performances as the years rolled by.

Mili was the leader in photographic application of the new system of lighting. His classical photographs are to this day models of imagination and technical skill. One of Mili's first accomplishments was a study in 1938 of Bobby Riggs for the then-fledgling *LIFE* Magazine. To *LIFE*'s viewers it provided the "pleasure of sudden wonder," and Mili was off and running as a successful *LIFE* photographer. The numerous flash pictures *LIFE* published in those early years helped to spread the fame of both Mili and the magazine.

Mili, less interested in the technology than Edgerton was, went on to perfect other photographic techniques that gave his pictures an unmistakable Mili identity, the hallmark of a distinguished and versatile photographer. A later accomplishment was to gain the confidence of Picasso and produce one of the best of all pictorial books on his sculptures.

Mili has been helpful to Edgerton and me in the preparation of this book, and we are privileged to include a group of his photographs.

The First Flash

In 1932, while I was editor of the *Technology Review*, Edgerton showed me some of his first and now-classic electronic flash still photographs, and I was delighted to publish a series of them month after month in the *Review*. Thanks to him, the *Review* was the first lay journal to publish a representative collection of his path-breaking photographs. At the same time a widening circle of audiences was finding his high-speed motion pictures absorbing, even astounding.

The widespread lay interest in this new strobe technology led me to conclude that the numerous technical papers Edgerton and his colleagues had been publishing on electronic flash systems needed to be supplemented by a book aimed at a general audience; so in 1938 we published a volume, *Flash: Seeing the Unseen With Ultra-High-Speed Photography*.

In the forty-one-year interval since *Flash* was published, many others have utilized the Edgerton system with high professional skill. Edgerton himself, together with his students, has continued to explore the absorbing world of fast motion, penetrating new domains of high-speed events in nature and refining his early transcriptions, papers, and articles in the technical press. *Flash* was the pioneer presentation in book form of Edgerton's stroboscopic technology and the remarkable photography it made possible. It is long out of print, and since its publication, "strobe" photography has yielded new harvests with its exploring eyes and demonstrated usefulness in areas not then imagined. It has given new capabilities to journalists, oceanographers, physicians, naturalists, explorers, archaeologists, and many others. During World War II, it contributed a new capability to the Allies' air reconnaissance; today it is becoming universally available to millions of amateur photographers.

In view of all of these advances and new applications of the strobe, an updated book is warranted. Many of us also feel that despite his brilliant lights, Edgerton's candle has been partly hidden under a bushel. For example, in the introduction to a 1976 exhibition of Edgerton photographs in Great Britain, Geoffrey Holt wrote, "Here is a photographer whose work we feel has been unjustly neglected ... Edgerton's work commands respect and interest as being the output of one of the masters of the optically conscious."

It is my hope that this album will stand as a definitive and up-to-date record of Edgerton's photographs, both the classic early ones of the 1930's and the more recent ones that have given sharper eyes to research specialists and explorers.

In recent years Edgerton has devoted much of his energy and imagination to underwater photography and its use for deep sea exploration, archaeology, and the study of marine life. As an aid to his underwater exploration, he has made inventive adaptations of long-understood sonar techniques and invented "pingers," which make it possible to "feel the bottom of the sea," and has adapted other sound devices. This new book encompasses these activities.

Finally, the book looks at the relation of strobe photography and the arts and provides a brief historical sketch of the development and use of the strobe light.

Stroboscope Breakthrough

While a graduate student in electrical engineering, Edgerton participated in a study of what electrical engineers call power system stability, the ability of synchronous generators and motors to stay in step after sudden disturbances such as lightning strokes hitting transmission lines. In this study he was trying to determine with accuracy the transient changes in the angular displacement of the rotor of a synchronous electric motor. These displacements in the speeding rotor could not be seen by the unaided eye. One remedy tried during this experiment was the use of mercury arc rectifiers to supply very fast reinforcement to a powerless generator. During the course of the experiment, Edgerton suddenly became aware of being able to see the rotating poles of the machine oscillate about a mid-position following a simulated system disturbance. The mercury arc rectifiers happened to be placed close enough to the generator for their flashes to illuminate its rotor stroboscopically. As described to me by his colleague, Harold Hazen, Edgerton's prepared mind immediately recognized the opportunity to develop strobscopes for the purpose of "stopping motion" and he proceeded to develop the first modern stroboscope, described in the May 1931 issue of the journal *Electrical Engineering*.

Two students, Kenneth Germeshausen and Herbert E. Grier, who were to become Edgerton's full partners, made important contributions in the early 1930's to the further development of strobe systems, notably many unique strobe lamps and circuits (Germeshausen) and electronic flash lamp equipment and circuits (Grier). The pioneering research of these three partners ultimately led to the development of the now widely used electronic speed flash system.

Around 1940, Edgerton and Germeshausen were invited to take their experimental stroboscopic high-speed motion-picture camera to the MGM Studio in Hollywood, California. A short assembled by Pete Smith, named "Quicker Than A Wink," resulted and received an Oscar for the best short of its year.

As many MIT graduate students and young staff members in engineering have done, the three investigators, moved by an entrepreneurial spirit, recognized that their academic work had industrial potential, and anyway they needed income. In 1947, they translated their informal partnership of thirteen years into a company and gave it their own names, "Edgerton, Germeshausen, and Grier," to manufacture equipment they had designed. This company slowly grew into a substantial electronic industrial organization. Later when corporate acronyms became stylish, the company renamed itself "EG&G," as it is known worldwide—and on the New York Stock Exchange.

Edgerton played a major part in launching and advancing the fortunes of this enterprise and he benefited munificently from its growth, but he refused to abandon his teaching and duties as a professor to become a full-time industrialist. Consequently, he left the management of the emerging enterprise to Germeshausen, Grier (all three founders are now retired), and others, and without missing a stride, continued unabated his full-time dedication to his students and his research, as in a memorable line of Keats, "forever panting and forever young."

As his novel and pioneering photographs awakened worldwide interest in strobe photography, Edgerton's laboratory became a mecca for people who wished to learn his techniques or to have him analyze rapid motion with his equipment. At the same time Edgerton himself took the initia-

Edgerton Makes a Discovery

One of the earliest (1931) high-speed pictures made with Dr. Edgerton's stroboscope. The motor rotor, on which the *S* and *N* are painted, was turning at a rate corresponding to a linear speed of 95 miles per hour when the exposure was made. It is Dr. Edgerton himself (age 28) who watches; the setting is an electrical engineering laboratory at the Massachusetts Institute of Technology.

J. Ralph Jackman

tive to identify, to see, and to understand what his flashing lights could reveal. He went to New Hampshire to photograph hummingbirds in flight; he went to the circus to capture the spectacular performances of bareback riders. I remember him photographing flying bats, baseball bats, and acrobats; a fancier of tumbler pigeons persuaded him to photograph their strange somersaults; great tennis players, baseball players, fencers, and golfers came to him for photographic analyses of their techniques. When the president of the DuPont Company, Crawford Greenewalt, became deeply engrossed in photographing hummingbirds, he turned to Edgerton for help. He later compiled his masterful portfolio of color photographs of hummingbirds in flight, a classic of natural history. Altogether it was a colorful parade of men and beasts into a new world of vision. In the meantime Edgerton had achieved dazzling visual imagery of running water, milk drops, and bullets.

The Explorer

Explorers, naturalists, and archaeologists as well as gifted professional photographers sought him out, and Edgerton became widely known for his contributions to oceanography. Captain Jacques-Yves Cousteau enlisted his help aboard the *Calypso,* searching under water for ancient sunken vessels and other archaeological artifacts in the Aegean Sea, and in exploring sea floors in the Mediterranean, the Atlantic, and Lake Titicaca in Bolivia. When Cousteau embarked on a search for the great hospital ship *Britannic,* sunk by a mine in Grecian waters in World War I, he again called on Edgerton for help. An American group led by Robert Rines ambitiously set out to find the Loch Ness monster—if there is one—and enlisted Edgerton. Rines commented wryly that Edgerton became "hooked" by the challenge, less by the search for the monster than by the exploration of circular structures on the floor of the Loch. Discovered by Klein and Finkelstein, these structures were reminiscent of Stonehenge.

Edgerton was a member of a group that set out to locate the sunken Civil War ship, *USS Monitor,* off Cape Hatteras, North Carolina. He knew of the capability of the side-scan sonar to explore large underwater areas for shipwrecks. Together with John Newton (Duke University Marine Laboratory), Robert Sheridan (University of Delaware), Gordon Watts (North Carolina archaeologist), and Fred Kelley (Duke University Marine Laboratory) aboard the *R/V Eastward* (1973), a successful search was made and reported by John Newton in the January 1975 issue of the *National Geographic* magazine.

In the search for the *Monitor* one of the Edgerton-type cameras was caught in the wreck and only recently has been recovered. It now hangs on the wall of "Strobe Alley" at MIT. Edgerton has remarked, "One of my favorite preoccupations is figuring out how to retrieve a long list of my cameras lost all over the world." He recommends that all cameras used in ocean depths be clearly marked with the name and address of the owner.

In 1969, a group of Soviet oceanographers invited Edgerton to sail with them from Britain to Boston via Dakar, Africa, on their elegant oceanographic vessel, the *Acadamik Kurchatov.* On this trip he took an MIT freshman, James Sholer (who spoke some Russian), and they demonstrated some of the Edgerton techniques by lowering an underwater camera four times into the rift valley in the mid-Atlantic. By the time the ship arrived in Boston Harbor, the Soviet oceanographers had gained ocean engineering knowledge from Edgerton, and he and the ebullient Soviet oceanographers were singing Russian and American songs. In these Russian songfests Harold taught them "She'll Be Comin' 'Round the Mountain when She Comes," and other lusty American ditties, and they in turn introduced Russian songs Harold says are too memorable to quote.

These are but a few vignettes of the explorer Edgerton. But even as he became world famous and a world traveler, his laboratory at MIT continued to be his base. It is off Strobe Alley, the corridor gallery of Edgerton photographs and a museum of artifacts and equipment that have been recovered from deep-sea waters. In a very real sense, this book is an illustrated monograph on the Edgerton gallery and museum. Even though he technically retired ten years ago, he is still a magnet for students, and he gladly teaches in both laboratory and classroom.

The Edgerton Strobe Technology

Beyond the time threshold of human vision lies a world of rapid motion which can be seen only with the aid of accessories that manipulate time as the microscope or telescope manipulates space. The unaided eye cannot see the beautiful coronetlike splashes of milk drops falling in a pool, the wings of a hovering hummingbird, or the effects of a bullet in flight. Not until the development of high-speed or single-flash photography (except by the use of sparks) did it become possible to see these rapid phenomena or visually to harness time to space.

High speed properly describes single-exposure cameras that take photographs with exposures shorter than 1/10,000 of a second, and motion-picture cameras that operate at speeds in excess of 300 pictures (frames) per second (the motion-picture camera normally operates at 16 to 24 frames per second). Most of the pictures in this book were taken with exposures shorter than 1/50,000 of a second, and motion pictures have been taken at a rate as high as 6,000 frames per second. In the laboratory at the Massachusetts Institute of Technology where most of the early pictures were made, an exposure of 1/100,000 of a second has become commonplace, and the unimaginable interval of 1/1,000,000 has become thoroughly domesticated and broken to harness as a useful part of the time scale. A millionth-of-a-second exposure! Let us represent one second by the distance—about 3,000 miles—from New York to San Francisco. Then 1/1,000,000 of a second is the distance across your living room, say about 15 feet. It makes that old expression "quick as a wink" seem absurdly gross, for high-speed pictures of the eye show that the average man or woman takes the interminably long time of 1/40 of a second to wink once. On the other hand, if we are to investigate the life of a rifle bullet traveling at 2,700 feet per second, or 1,800 miles per hour, a millionth of a second is not too brief an exposure.

Such speeds are necessary, then, to capture ultra-rapid motion. As an illustration, take a golf ball zooming away after an average drive. An ordinary camera, with its shutter set at 1/1,000 of a second, would photograph the traveling golf ball as a blur about 5 centimeters long. The exposure time must be less than 1/100,000 of a second in order to produce a sharp, clear photograph of the ball. So far no practical mechanical camera shutter has yet been built to perform as fast as this. Furthermore, if it could be constructed, there would still remain the almost insuperable problem of obtaining sufficient light to expose the film in so short a time as 1/100,000 of a second.

To take the kind of high-speed pictures shown in this book, electrical control of the illumination replaces shutters. The light is turned on only when a picture is to be exposed, in a fashion similar to taking photographs with flashbulbs, except that the duration of the flash from a flashbulb is some 2,500 times longer than the 1/100,000 of a second maximum exposure necessary for "freezing" the golf ball. The flash of the light is substituted for the opening and closing of a shutter. This does not mean that an ordinary camera cannot be used to take high-speed pictures (all the pictures in this book were taken with standard cameras); it means that the exposure is made by a light flashing on and ott more quickly than any shutter can open and close.

Professor Edgerton and his associates, Germeshausen and Grier, designed lighting equipment for providing either a single flash for taking a single "still" picture, or a series of flashes with a predetermined interval of time between them for taking multiple-exposure photographs and motion pictures. Whether there is a single flash or a succession of flashes, the light is produced by an electric spark—miniature lightning—inside a gas-filled lamp. Electricity flows into a kind of electrical reservoir known as a condenser, and when the reservoir is full, it overflows at the desired instant to produce a brilliant flash inside the lamp. Electrical controls make it possible to govern very accurately the time between flashes and the exact moment of the flashes. To take one of the bullet pictures in the book required a single flash lasting about 1/1,000,000 of a second. A pulsing light is technically called a stroboscope, meaning roughly "whirling watcher," although the whole family of bright, fast flashing lights, including those used with amateur cameras, now bears the generalized name of *strobe*.

The flashing lamp, which is the heart of the Edgerton system of speed photography, has another important quality. It produces a light of great actinic intensity and is many times more brilliant than sunlight, as of course it must be if the exposures used are to yield good pictures. Even though the eye seeing it is unaware of unusual brightness, the instantaneous intensity of each flash exceeds the light of approximately 40,000 50-watt bulbs, such as those used in household lighting. This intense illumination allows photographic exposures to be made at small apertures, *f*11 or less. With this equipment the taking of speed photographs becomes relatively simple. Any standard camera and lens equipment may be used to get effective exposures up to 1/1,000,000 of a second without special fast emulsions, although they are useful. Moreover, the pictures obtained are true reflected-light photographs with good depth and detail.

The method of taking "still" pictures of moving objects with a single flash of light is not new; it is almost as old as photography itself. William Henry Fox Talbot, who shares with Daguerre the honor of taking the first photographs ever made and discovered the calotype process and the negative-positive technique, patented a method of instantaneous photography in 1851. In a darkened room he focused a camera on a rapidly revolving disk, and with an electric spark produced by the discharge of a battery of Leyden jars, an early form of electrical capacitor, he obtained an unblurred picture of a clipping from the *London Times* which was attached to the revolving disk.

While Talbot was the first to demonstrate high-speed photography by use of electric sparks, five years earlier Charles Wheatstone "froze" motion by using the bright flash of an electric spark in a dark room to "stop" rapidly moving objects for a fraction of a second. Unfortunately, photography was not yet available to record the demonstration.

After Talbot, spark photography was developed to a high state of perfection by Mach (of Mach number fame), Cranz, Salcher, Boys, and others, particularly for photographing bullets or other fast-moving objects in silhouette. Silhouette spark pictures of bullets yield inportant information by showing the shadows of the sound waves produced. These waves become visible because changes in the density of the air vary its refractive index, and this variation is recorded by silhouette photographs.

By utilizing electronic devices, Edgerton has been able to substitute a tube or lamp with a control circuit for the electric spark. This book presents striking examples of silhouette pictures taken by the magic lights of Edgerton with his ingenious way of triggering them.

A common illustration of stroboscopic sight is the use of a flashing light to read letters painted on the blades of a revolving electric fan, similar to Talbot's experiment. Suppose the fan is turning 18 revolutions per second. Seen in ordinary light it looks like a whirling disk because the eye is too slow to pick out the separate blades, just as the ordinary camera blurs the golf ball. But by examining the rotating fan under stroboscopic light flashing at the same rate (18 times a second), an observer sees the fan apparently still and can read the letters.

Not only can the fan be made to "stand still"; it can be viewed in slow motion, although its actual speed is still 18 revolutions per second. By regulating the stroboscopic light to 17 flashes per second, the fan will appear to be turning only one revolution per second, the difference between 18 and 17. By speeding up the flashing rate of the light beyond the speed of the fan blades, the fan will appear to turn slowly backward. Motion can thus appear juggled, slowed, frozen, or reversed by the stroboscope.

The stroboscope in its original form was not a flashing light; it was a disk, with alternate open and closed areas, which, when revolving, gave an intermittent view of a moving object or series of objects. The reader can easily construct this simple mechanical "whirling watcher" himself. Cut a disk from a large piece of cardboard and near its rim cut a series of slots equally spaced and large enough to look through when held close to the eye. From the center of the disk draw radial lines, extending from the center like the spokes of a wheel, one line for each slot. Then mount the disk so that it may be held vertically and revolved rapidly. Now hold the disk in front of a mirror so that you can look through a slot and see the image of the radial lines in the mirror. When you set the disk spinning and look through the successive openings provided by the slots, you will see in the mirror not the blur you would normally see but the image of each radial line distinct and standing still.

The explanation of this illusion is simple if we recall that our eyes have a quality known as persistence of vision or *retinal lag*. We continue to see an object, line, or light for about one-tenth of a second or less after it has moved away. The four blades of the revolving electric fan, for example, appear as a solid disk because the image of each blade persists in the eye after the blade has moved by.

When we look through the slots of the turning cardboard disk, each slot as it passes the eye gives us a glimpse of the image of the disk in the mirror, but the glimpse is so short that the image does not have time to move enough to show any motion. The lines appear to be standing still. When the second slot passes the eye, giving a similar successive view, the first view still persists in the eye and the second merges with it into one single impression. The result is an image that continues to stand still.

With this same device we can make the radial lines move slowly forward or slowly backward, just as the blades of the electric fan move when the speed of the flashing light is changed, and for the same fundamental reason. If we draw more radial lines than there are slots on the disk, we get the first effect of slow forward motion. If there are fewer lines than slots, we see the second effect of slow backward motion. Herein lies the explanation of that illusion we have all seen in the movies — a bicycle wheel turning backward when the bike is clearly moving forward. The intermittent action of the motion-picture camera shutter corresponds to the eye looking through the slots on the disk. If the camera takes pictures at a rate faster than the bicycle wheel turns, or if the lines are less numerous than the slots on the disk, we receive a view of the next spoke or radial line slightly before it has arrived at the spot where the first was seen, giving the illusion that the first spoke or line has moved backward.

In this simple slotted disk lies the whole secret of the stroboscope, of seeing rapidly moving objects standing still or in slow motion. A moving image is seen so briefly through a slot, shutter, or by the flash of a light that it appears not to move, and this image of nonmotion persists in the eye until the next image replaces it. Of course it is this physiological quality of persistence of vision that makes the motion picture possible. Thus the slotted disk — the first stroboscope — was the progenitor of the cinema.

The birth of the stroboscope in 1832 deserves to be recalled. Like many inventions, it was developed independently by two men, Plateau of Ghent and Stampfer of Vienna. It seems both were inspired by some investigations of Michael Faraday. He, in turn, was probably prompted to investigate observations associated with persistence of vision made by Peter Mark Roget, the British physician who compiled the famous *Thesaurus of English Words and Phrases*. There were versatile men in those days! Even though Stampfer was a month or so behind Plateau in announcing his invention, the name *stroboscope*, which he gave to the device, has persisted rather than the name Plateau coined for his own apparatus, the more formidable *phenakistoscope*. Plateau made many original contributions to physiological optics, as well as to molecular physics, but the last forty years of his life were spent in complete blindness as a result of gazing too long at the sun. He lost his own sight, but he helped to give a new vision to mankind.

Not long after the invention of the stroboscope, viewing slots were adapted for interrupting a beam of light, thus mechanically providing the periodic flashes later to be produced electrically.

Dr. Edgerton's high-speed motion-picture photography, because it uses a controlled, pulsing light, is called *stroboscopic photography*. His contribution has been not only to put the camera and stroboscope into an effective double harness but to develop a unique stroboscope that can be controlled and accurately timed and that supplies in readily usable form the brilliant light necessary for photography. He experimented with a number of different light sources, and out of this work came a battery of magic lights for many specialized uses.

Edgerton also overcame the difficulty of flashing the light

at the right time to photograph a bullet at a desired position or a tennis ball when it was flattened by the racket. He accomplished this control of the starting time in a variety of ways, depending upon the subject being photographed. Frequently he used a simple electrical contact actuated by the object itself. For other photographs he let sound trip his circuit with a microphone, or he used a signal produced by the interruption of a beam of light from a photoelectric cell. Much of the success of his pictures resulted from his ingenuity in devising these methods for accurately timing the flash that exposes the film.

While no special camera is needed for taking high-speed still pictures, high-speed motion pictures require a camera that moves the film fast enough and without any intermittent action. Naturally the faster the action that is being photographed, the faster the film must be driven. It is impractical, because of mechanical limitations, to drive the usual motion-picture camera at speeds in excess of about ten times normal, or 240 frames per second. Conventional pictures taken at an increased rate are familiar to all; projected at ordinary speeds, they are slow-motion moving pictures.

To obtain really high-speed pictures, and thereby ultra-slow-motion projection, the film must be whipped past the lens at a speed much greater than 240 frames per second. To gain such speed, Edgerton has utilized special cameras that avoid the intermittent film-fed mechanism entirely and instead use a continuously moving film mechanism synchronized with the flashing light. Each time the film has moved the distance occupied by one frame (or by half a frame, if desired), the subject is illuminated by the stroboscope and the film is exposed. While the film is running, the camera lens remains wide open. The flashing light replaces the shutter action and is so instantaneous that the racing film has no blurring effect. Normal illumination, such as that encountered indoors, is insufficient to fog the film in these cameras because the film passes the lens so rapidly. When taking 2,000 standard size pictures a second, the speed of the film approaches 85 miles an hour, which means 7,500 feet per minute, or 125 feet per second.

One of the most difficult problems that Dr. Edgerton had to solve in making motion pictures at these speeds was to frame the pictures properly so that they could be projected in standard motion-picture equipment. Experts opined that it could not be done. That it has been done will be verified by the many thousands of persons who have viewed the film produced by Edgerton, entitled "Seeing the Unseen." What Edgerton did was to harness the electrical circuit controlling the light to an accurately constructed contactor driven by the sprocket moving the film. This provided perfect synchronism between the racing film and the flashing light.

The principal limitation of this stroboscopic type of high-speed camera is its inability to record self-luminous subjects such as electric sparks and explosions; but several different systems have been developed to photograph such phenomena. Instead of illuminating the subject for a sufficiently short length of time to avoid blurring, as Edgerton's method does, these systems employ moving optical arrangements which hold the subject stationary with respect to a continuously moving film. In other words, by moving a lens, mirror, prism or slit so that the image moves in the direction of the film and with the same velocity, it is possible to take as many as 120,000 pictures a second, provided the object being photographed produces sufficient light itself or else can be adequately illuminated. Cameras of this general type were constructed by Jenkins, Thun, Magnan, Suhara, Tuttle, Prince Rankin, Waddell, and Shoberg.

In addition to its capacity to record vagrant and rapid motion, stroboscopic photography is useful for measurement and analysis. Since the interval between flashes may be predetermined, the observer is able to record an action as a function of time. Likewise, the distance, velocity, and acceleration of the moving object photographed may be read and calculated from the pictures. To return to the golf ball pictures, it is possible to determine from them the velocities of the ball and the club, as well as the spin of the ball, the angle of departure, and the twist of the club head after it has hit the ball.

The method of measuring the velocities is this: if two high-speed exposures of a moving object are taken at some small interval apart, the object will appear to have moved a short distance. The displacement or motion of the object divided by the interval of time is the average velocity during the interval between the two photographs. Thus if the ball moves two inches in 1/1,000 of a second, then its velocity is 2,000 inches (167 feet) a second. In the multiple-image photographs of golf strokes in the sports section this measurement may be made easily from one photograph. Each image of the club and ball follows the previous one by 1/100 of a second or some other small interval. Knowing this, one readily obtains the duration of the stroke, or any part of it, and the speed of the ball. This ability makes stroboscopic photography a powerful and widely useful tool in scientific research whether that research deals with abstruse laboratory phenomena, the behavior of machinery in a factory, the flight of hummingbirds, or the mechanics of pitching a baseball.

The single-flash strobe has been put to work by the commercial photographer, who uses it to make portraits as well as action pictures. As Fabian Bachrach, the well-known portraitist once wrote to Edgerton, "...you have made it possible to capture spontaneous actions and emotions never before attainable." It gives the photographer a further range, a new mastery of light, and a greater freedom to pursue aesthetic objectives. It also serves as a substitute for flash bulbs. The Edgerton lights will flash repeatedly without having to be replaced, although they are not so portable as flash bulbs. The slotted disk of blind Dr. Plateau in its modern electronic version has taken its place in the workaday world.

Strobe Photography and Art

In her recent book, *On Photography*, which expresses a deprecatory attitude toward all photography, Susan Sontag observes that "Whatever the claims for photography as a form of personal expression on a par with painting, it remains true that its originality is inextricably linked to the powers of the machine: no one can deny the informativeness and formal beauty of many photographs made possible by the steady growth of these powers, like Harold Edgerton's high-speed photographs of a bullet hitting its target, of the swirls and eddies of a tennis stroke...."

True enough, but Edgerton's photographs frequently transcend the "powers of the machine." As Geoffrey W. Holt has written, "his photographs are remarkable because even at a millionth of a second, his work has its own identity, as recognisable as those which distinguish more conventional photographers. Such is the power of these photographs, they have, within his own lifetime, become classic visual images."

Those of us who observe Edgerton at work cannot fail to note his preoccupation with craftsmanship and aesthetic aims. He clearly seeks to achieve a picture that represents his human judgment of what is right and what embodies his own identity. He seeks to master a mechanical or electronic system so that it achieves the imprint of his own human decision and permits the best possible correlation between meaning and expression. Remember that in the split-second realm in which he works, this is the more difficult—the best evidence that many of his photographs appropriately fall in the category of art. If for no other reason than that they are clearly artistic revelations, they "sign themselves" as the work of Edgerton and they introduce us to the beauties of that new land of vision so strikingly described in Gyorgy Kepes's books, *Language of Vision* and *The New Landscape*.

In making these observations, I fully recognize that as a craftsman, Edgerton is primarily interested in analyses and hitherto undiscovered information. While he hopes that his pictures please as well as instruct, as many of them do, he does not seek a style charged with mystical inner meaning, metaphor, or emotion, as do some contemporary photographers.

It would be misleading, then, not to recognize that Edgerton is first and foremost a scientist and engineer. When he devised the modern stroboscope, he was not trying to become a photographer; he was trying to find a way to solve an engineering problem. Throughout his career he has invented new photographic systems as tools to solve problems. In this work his photography is sometimes secondary to the information he seeks. Photography buffs may be surprised at the apparently poor photographic quality of some of the pictures that appear in the latter part of this book, the earlier parts of which contain so many photographs that stand alone by their clarity. There are also other Edgerton photographs in these pages that are seemingly bizarre and some that seem unintelligible or amateurish at first glance. These photographs,

The Teacher and the Student

A Pictorial Statement of Professor Edgerton's Art and Love of Teaching

The student asked Professor Edgerton for a print of the classic strobe-light coronet photograph. Instead of giving him the print, Doc replied, "We've never gotten a perfect crown of the splash. If you will take time to learn how to work the camera, and if you will help me, we'll try one together; the sooner the better."

Gjon Mili

Until Velásquez painted his great *Las Hilanderas (The Spinners)*, about 1660, no previous painters had captured the stroboscopic effect. In reproducing a turning wheel, they were accustomed to painting each spoke as though it were standing still. In the segment of the Velásquez painting reproduced, the motion of the spinning wheel is shown as it appears in reality with "invisible spokes and diffuse circular formation around the spindle."[4] This famous painting is in the Prado Museum in Madrid, Spain.

however, are an important part of the Edgerton achievement because they contain valuable and frequently unique information, or solve problems by providing a vision of scenes and objects otherwise unseeable. Photographs taken seven miles down in murky ocean water may not have the clarity of Edward Weston, Paul Strand, or Ansel Adams masterpieces, but to an oceanographer or an ichthyologist they may be beautiful because they contain information never before known and images never before seen. It is also important to emphasize that many of the aesthetically lovely photographs in the earlier pages of this book, which appeal to our sense of beauty, also contain new and useful information not always apparent to the layman but made uniquely available by electronic flash.

The Futurist Painters

Edgerton recoils from any discussion of the artistic values of his photographs, but there can be no question that the "magic light," whether it be spark or strobe, or single or multiple flashes, has influenced painting, or at times verified the theories of avant-garde artists. Around 1912 a school of painters, mainly in Italy, who called themselves Futurists, published manifestos expressing radical ideas about both politics and art. Their doctrine proclaimed that motion destroyed form. Marcel Duchamp stated, according to Beaumont Newhall, "that when he was painting his famous *Nude Descending a Staircase*, art circles in Paris were stimulated by stroboscopic and multiple-exposure high-speed photographs."[2] Perhaps the best known Futurist attempts to give static expression of motion by a painting technique resembling multiple-exposure photographs was Giacomo Balla's painting, *Dynamism of a Dog on a Leash (Leash in Motion)*. Antonio Giulio Bragaglia, a Futurist contemporary of Balla's, made photographic time exposures of moving objects and published them in his book *Fotodinamismo Futurista*.[3] It is interesting to speculate on how a modern strobe would have influenced the Futurists' doctrines!

Independently of the Futurists, the great constructivist, Naum Garbo, experimented in capturing motion in photographs, as for example in his beautiful *Kinetic Construction: Vibrating Spring* (1920). More recently (1956) the distinguished American photographer Harry Callahan published a delicate multiple-exposure photograph of a tree.

In his classic text *The New Landscape in Art and Science* (1956), Gyorgy Kepes gives comprehensive attention to the syncretic relationship of science to the arts, a relationship that in the broadest way has especially influenced modern art. Edgerton's strobe photography is but one example of this influence, but a powerful one.

On the occasion of Edgerton's retirement as Institute Professor at MIT, Ansel Adams wrote him: "In lecturing and teaching and photographing over many years, trying to get under the scaly epidermis of 'reality' to reach the essences of life and the creative spirit, I have often returned to your magnificent work for confidence and enhanced perception."

John Szarkowski, director of the Department of Photography at the Museum of Modern Art, also wrote: "Certainly one of the greatest pleasures for me in putting together the *Once Invisible* exhibition [at the Museum of Modern Art] was that of meeting finally the man whose pictures have hung on our Museum's walls almost as consistently as Picasso's have—the coronet of the milk drop, the swing of Don Budge, the foot of Wes Fesler—these and others must be as familiar to many of our visitors as the Guernica. But as fine as the pictures are, they are certainly not more interesting than the mind of the man that made them. . . . For the privilege of knowing the man as well as one corner of his richly various work I am lastingly grateful."

A Biographical Note on Papa Flash

Harold Edgerton lived in a variety of environments when he was growing up. He was born in Fremont, Nebraska, in 1903, where his father, Frank, was principal of the high school and coach of the football team. A couple of years after Harold's birth, the family moved to Washington, where his father served for a period as secretary to a senator and also worked as a reporter for the *Washington Times*. While in Washington, father Edgerton moonlighted to study law.

Washington, however, could not compete with the attractions of Nebraska, so the family moved back to their

Marcel Duchamp, whose *Nude Descending a Staircase,* No. 1 (1911), was a sensation at the Armory show in New York, is credited with saying, "My aim was a static representation of movement—a static composition of indications of various positions taken by a form in movement. . . ."

Reproduced by permission of the Philadelphia Museum of Art: The Louise and Walter Arensberg Collection.

14–15

Dynamism of a Dog on a Leash, painted by Giacomo Balla (1912). Oil on canvas, 35⅜″ x 43¼″. In accord with the doctrine of the Futurists, Balla sought to represent motion through the static medium of a painting. The Futurists, in pursuing their aim to display "the beauty of speed," were influenced by the crude high-speed photographs of the day.

Reproduced by permission of the Albright-Knox Art Gallery, Buffalo, New York, courtesy of George F. Goodyear and The Buffalo Fine Arts Academy. Bequest of A. Conger Goodyear to George F. Goodyear, life interest, and Albright-Knox Art Gallery.

beloved state, residing in Lincoln and on the Winnebago Indian Reservation. In all of these settings Mr. Edgerton practiced law, and when a law firm in Aurora, Nebraska, invited him to join them, he accepted, and the remainder of Harold's youth was spent in Aurora.

His experience in moving from one environment to another reminds me of my own youth, when I spent happy days successively in six small towns in North and South Carolina and Georgia. Growing up in a variety of small towns, each different and each requiring a special adjustment, was a rare educational experience which has stood both Edgerton and me in good stead.

Volta Torrey, a fellow citizen of Aurora, later an editor at MIT, and now with NASA, has described his fellow Aurorian of those days:

"The day that I first saw Harold he was wearing spurs and climbing a pole to help my father find out why the light had gone out in a movie house in Aurora, Nebraska. He came down grinning and said, 'cracked insulator.'

"Success often swells men's egos but left his untouched. Boredom frequently sneaks up on some of us, but he has eluded it. Everyone's door has been ajar for him, because merriment follows him like a shadow softened by kindliness."

After receiving a bachelor's degree in electrical engineering at the University of Nebraska, Edgerton spent a year at the General Electric Company in Schenectady, New York, taking the test course. He next came to MIT, where he won both a master's degree and a doctorate in electrical engineering and where his talents and high competence ultimately led to his appointment to the faculty and to his achievement of the highest rank at the Institute, that of Institute Professor.

Esther Garrett also came to Boston, to study voice at the Conservatory of Music, and in 1928 they were married.

As Harold Hazen, Edgerton's colleague and for years head of the MIT Electrical Engineering Department, was to write:

> . . . Harold is known by a number of people who have been exposed to his "Let's try it and find out what happens," and "Let's get going." They have felt his infectious enthusiasm and his friendly warmth that are virtually irresistible to a freshman or a corporation executive, or a denizen of the deep. . . . I doubt that anyone, and I include Harold himself, grasps the range and depth of his influence and inspiration. The impact of his curiosity, enthusiasm, down-to earth realism, superb experimental ability, and sheer productivity in new, untried areas, his outgoing spirit, his always-warm, simple friendliness and generosity, the impact of his personality—these are overwhelming as expressed by his students, his colleagues, and his admirers from all over the world.

I had the heady experience of reading over four hundred letters from all over the world that were written, as invited by Hazen, by Edgerton fans on the occasion of his retirement. As readers cannot fail to miss, these encomiums have helped shape and enrich this biographical sketch of Harold.

To these tributes should be added that in 1973 Edgerton received the National Medal of Science "for his vision and creativity in pioneering the field of stroboscopic photography and for his many inventions of instruments for exploring the great depths of the ocean."

When the Sea Grant Program at MIT recently acquired a vessel for oceanographic research, they asked Edgerton's permission to name it *The Edgerton*. Prompted by the same admiration, I have written these words about him with pleasure and devotion.

A sense of privilege in knowing Harold Edgerton is endemic among those who have been or continue to be his students, for Edgerton above all is a man who finds his deepest satisfaction in helping young men and women to grow and to solve problems.

For years he and Mrs. Edgerton held an annual party for his students at their home. There would be a flashing strobe at the door, an informal feast, and then Harold would bring out his guitar and the evening would be capped with a songfest.

For all those who delight in learning, in building and inventing, the strobe light is always at the Edgerton door, and for all who have had the privilege of association with him, there is a songfest in their hearts.

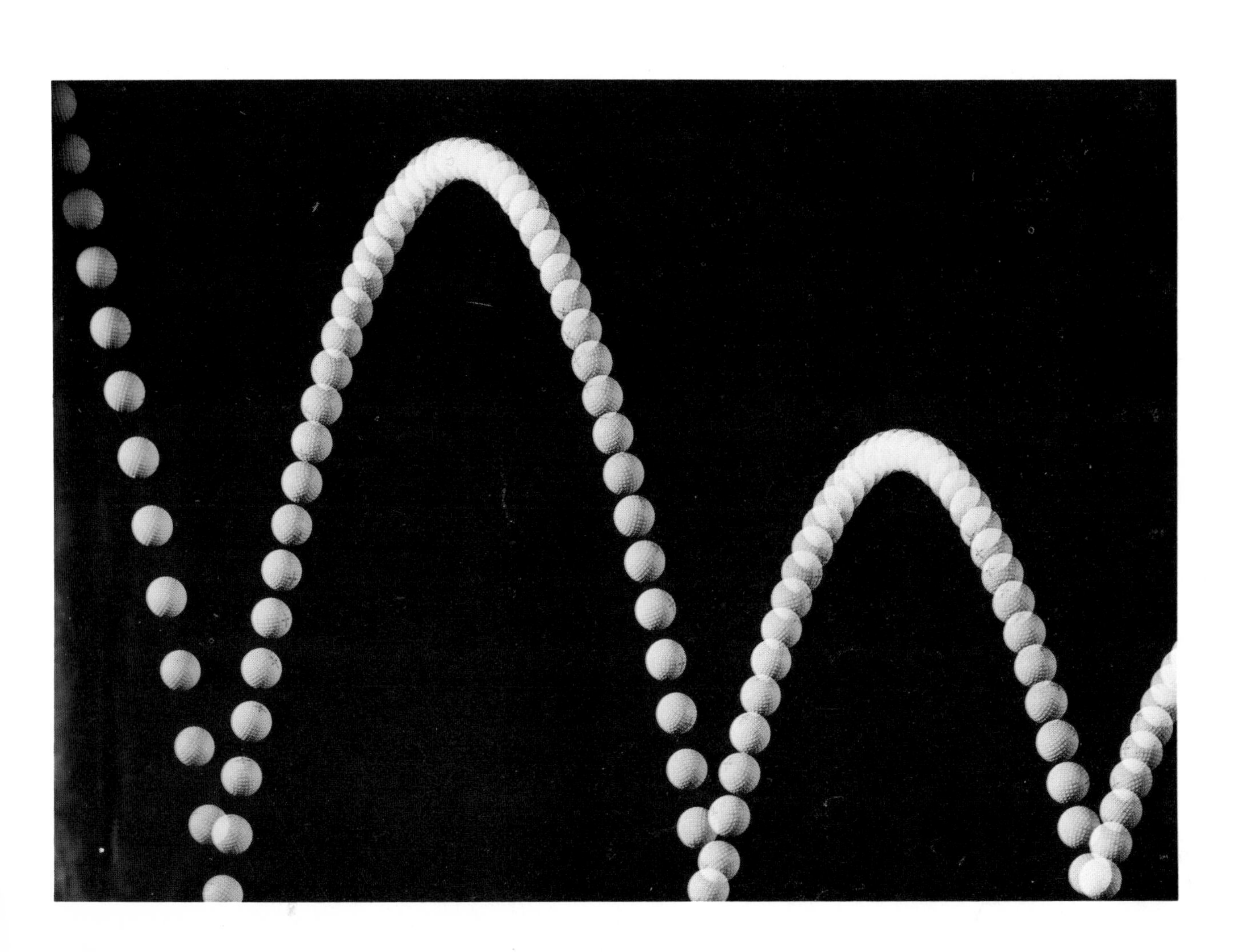

Drops and Splashes

In 1908, A. M. Worthington, professor of physics at the Royal Naval Engineering College, Devonport, England, published *A Study of Splashes*. From his fourteen-year investigation, he wished "to share... some of the delight that I have myself felt, in contemplating the exquisite forms that the camera has revealed, and in watching the progress of a multitude of events, compressed indeed within the limits of a few hundredths of a second, but none the less orderly and inevitable...."[5] Professor Worthington's photographs were the more remarkable for having been made in darkness with an electric spark, and he obtained many valuable experimental facts about surface tension and the changes of form that take place in the bounding surface of a liquid.

The pictures on the following pages represent a continuation of this vanguard investigation. "It would be an immense convenience," said he in his conclusion, "if we could use a kinematograph and watch such a splash in broad daylight, without the troublesome necessity of providing darkness and an electric spark. But the difficulties of contriving an exposure of the whole lens short enough to prevent blurring, either from the motion of the object or from that of the rapidly shifting sensitive film, are very great, and anyone who may be able to overcome them satisfactorily will find a multitude of applications awaiting his invention."[6]

Edgerton achieved this invention in his stroboscope. Under the revealing light of his stroboscope, it is now possible to see the exquisite splash formations with the unaided eye, to photograph them in daylight, and to take motion pictures for ultraslow projection.

In looking at the pictures of drops and splashes, two important principles should be kept in mind. First, the behavior of liquids is affected by surface tension. The surface layers of any liquid act like a stretched skin or membrane (a drumhead, for example) that is always trying to contract and diminish its area. Second, a spout or column of liquid, beyond a certain length in relation to its diameter, is unstable and tends to break down into a series of equidistant drops. As these drops are formed, they are joined together by narrow necks of liquid which in turn break up into smaller drops. A striking illustration may be seen in an ocean wave. At first the wave may have a smooth cylindrical edge. Then as the crest curls over, the edge becomes comblike, each tooth a jet that quickly breaks up into drops. This "principle of segmentation," be it noted, was first formulated by the blind physicist Plateau, who constructed the first stroboscope.

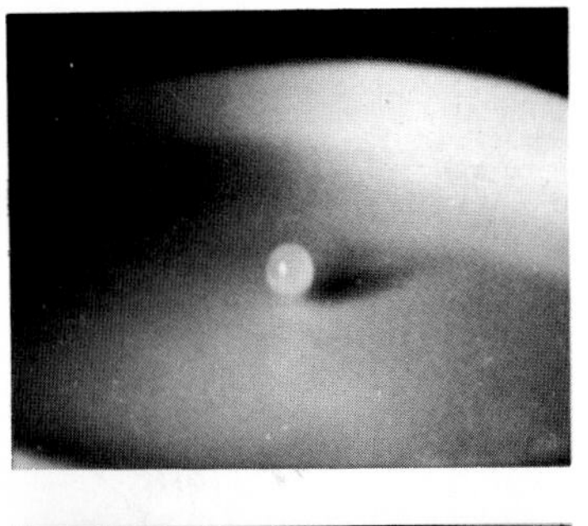
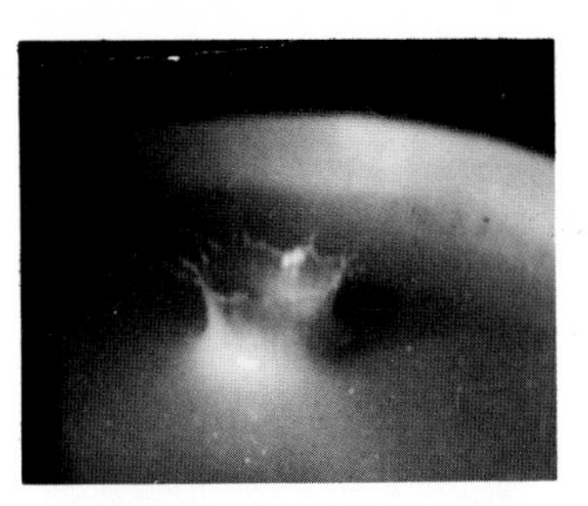
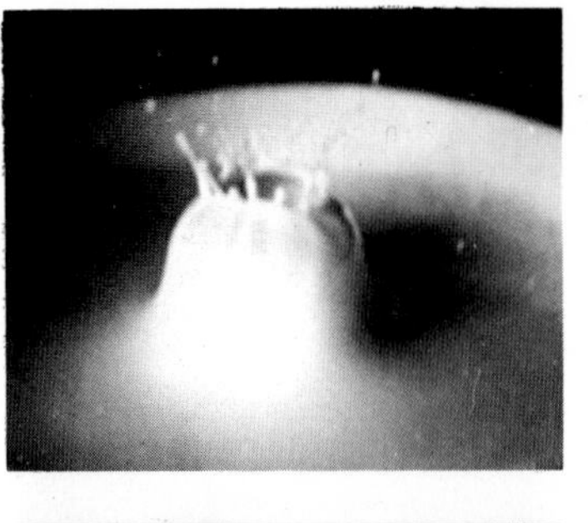
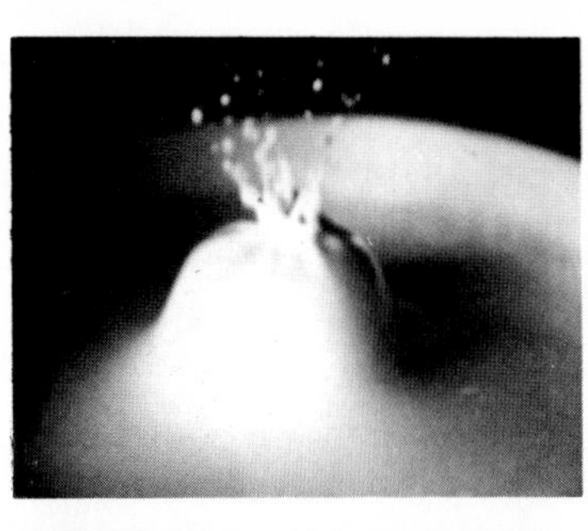
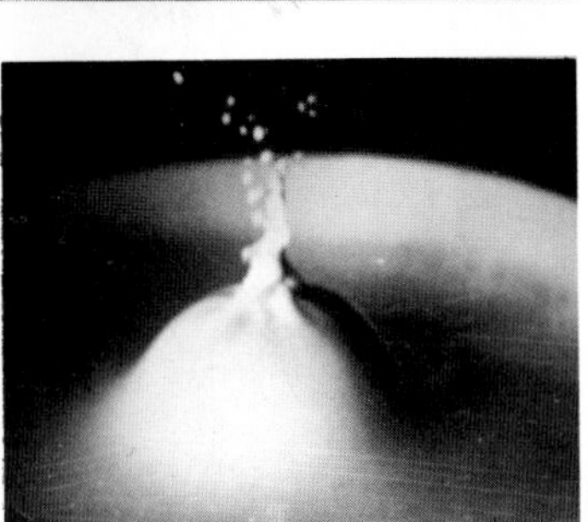
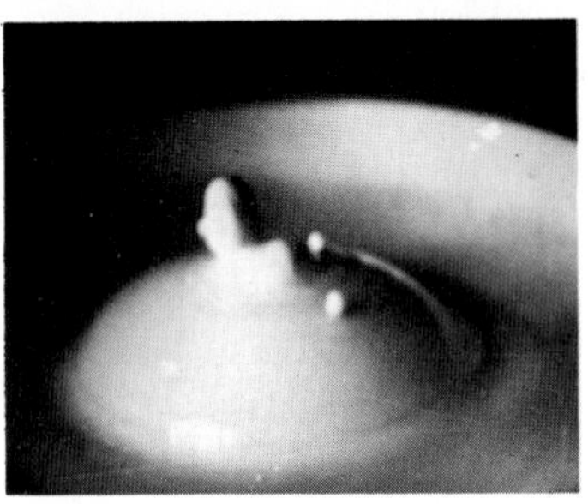
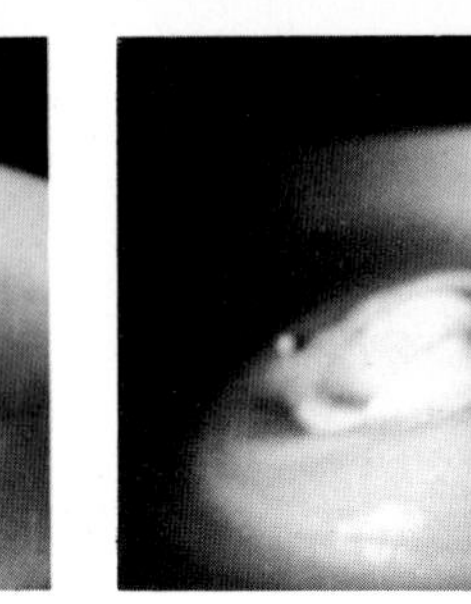
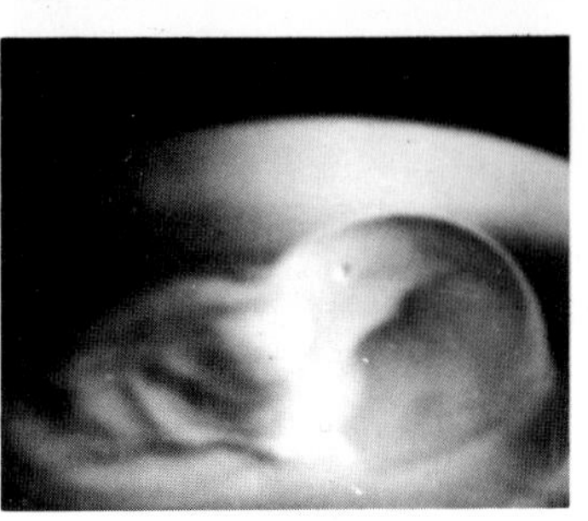
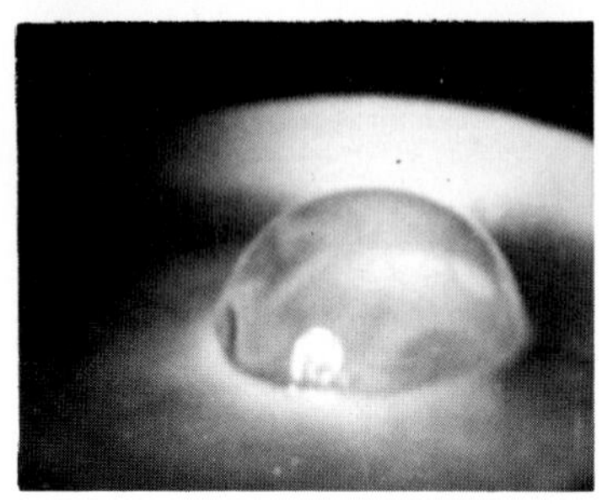

Birth of a Bubble
A Drama of Surface Tension in Ten Scenes

When raindrops splash into a pool of water, they frequently form large bubbles that float on the surface for a brief span. Here this familiar phenomenon—the parturition of a bubble—is captured in detail in a sequence of ten pictures. Instead of water, milk is used because the photographic emulsion shows it more clearly.

In picture 1, the drop is plummeting downward toward the surface of the pool of milk. Picture 2 catches the delicate crownlike formation of the splash, and pictures 3, 4, and 5 show the forces of surface tension, like the strings of a tobacco bag, acting to close the crown at the top and to entrap air within. In picture 6 comes the climax as the bubble, with an eruption of droplets, closes tight; in 7, 8, and 9 may be seen its contortions as it overcomes, by a narrow margin, the disruptive forces set up by the splash, and finally, in 10, the shimmering bubble stands full-bodied and complete.

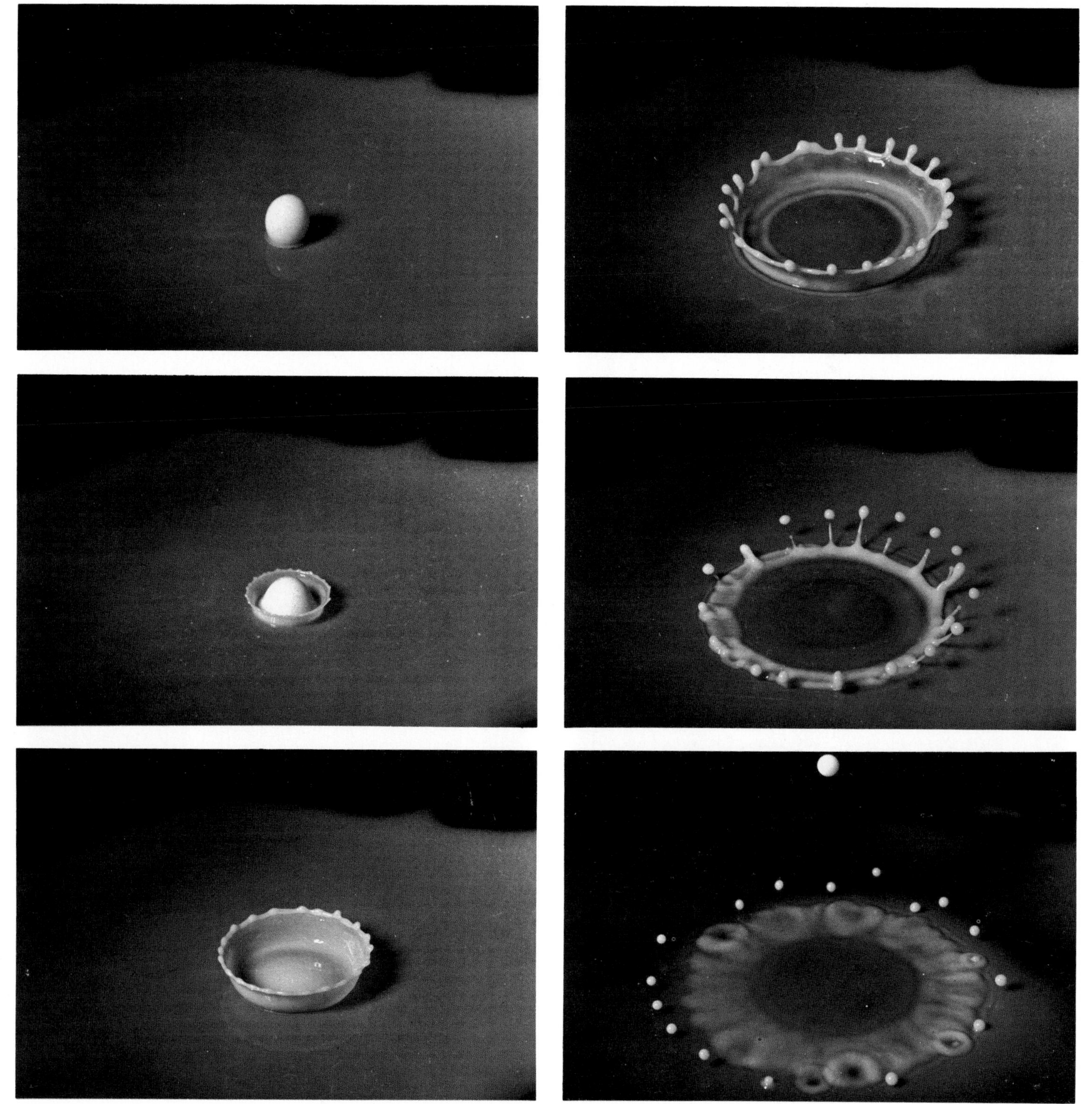

Drop of Milk Splashing on a Plate

Showing the crownlike formation of the splash

Drop Falling into Cup of Milk

Showing the formation and segmentation of the spout

Formation of a Drop

Milk dropping from a glass tube. Each picture is of a different drop; each was taken under the same conditions but at a different distance from the origin. A small drop follows the larger and during its oscillations takes the unexpected shape revealed in picture 5. Picture 8 shows a drop that has fallen six or eight feet. During this fall the oscillations have ceased, and the forces of windage and surface tension have gradually reached a balance, resulting in the stable shape shown. The drop in picture 9 has fallen 14 feet and flattened slightly but has the same contour as the drop in 8.

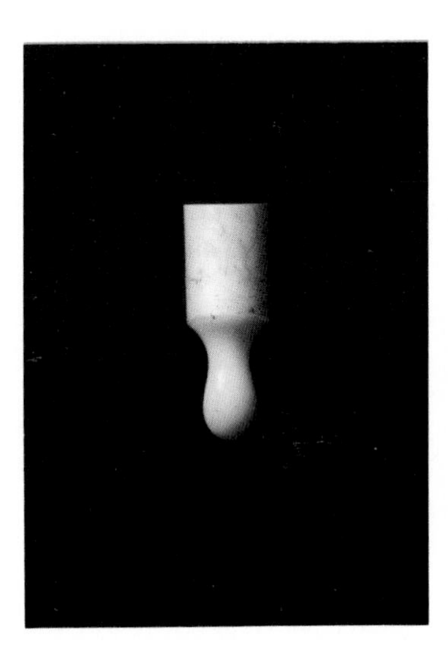

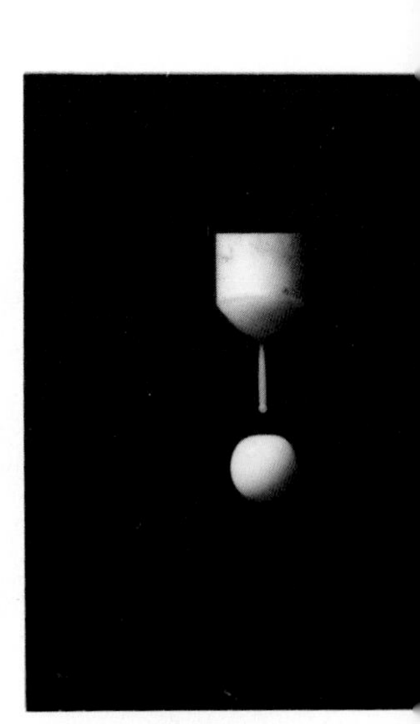

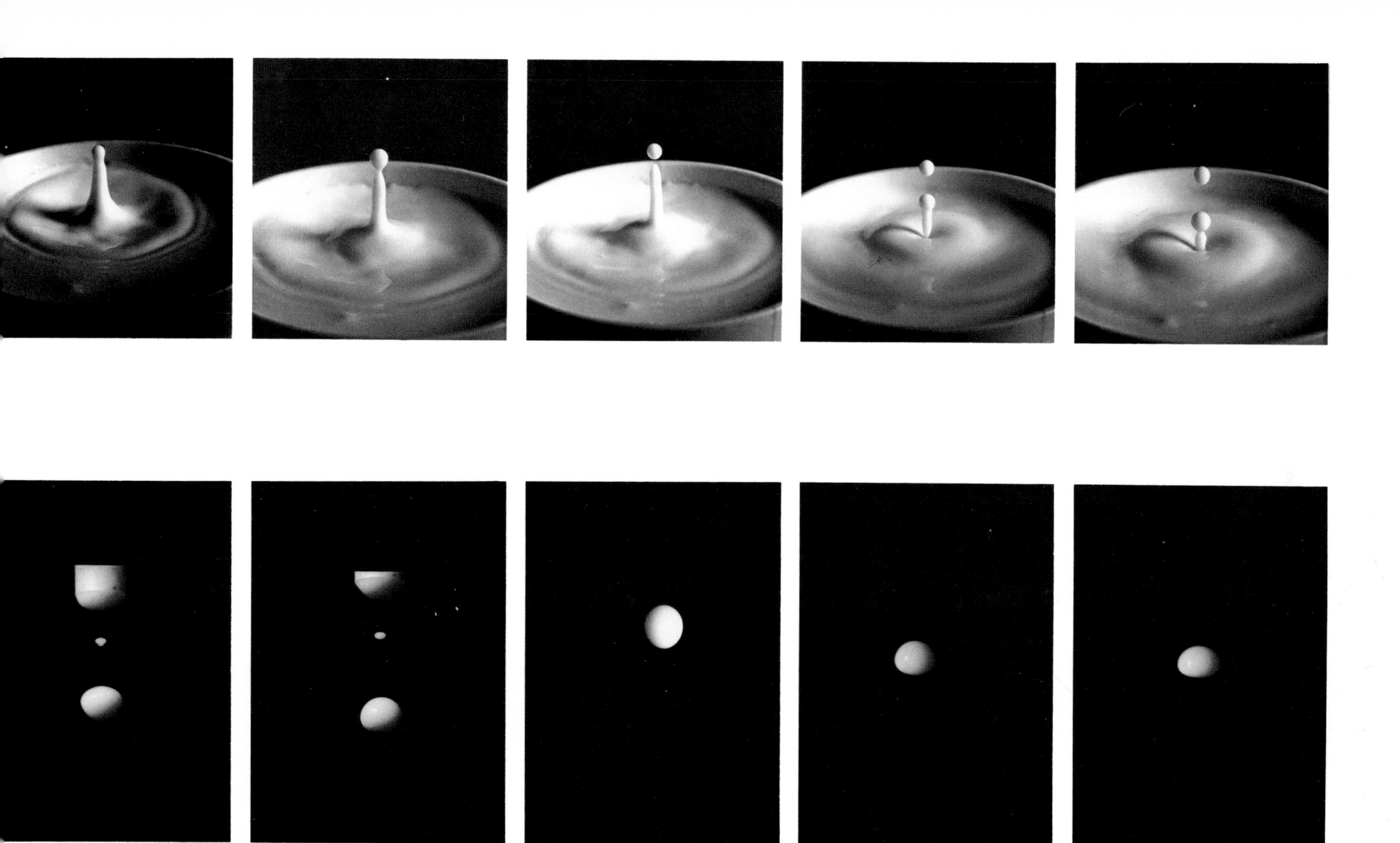

Milk Drops

Thicker layer of milk

Ronald Bucchino

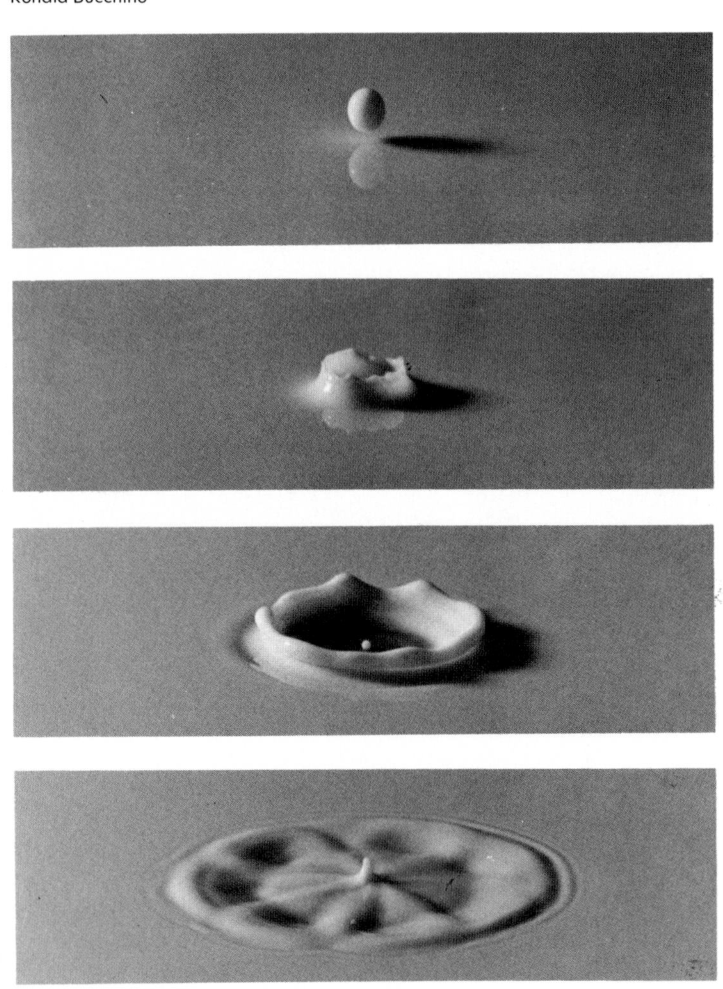

Spout

Close-up of spout shows drop breaking off.

Shape of a Long-Distance Drop

The water drop above had fallen eight stories down an elevator shaft when caught by the high-speed camera. The drop, pulsating as it fell, did not always fall straight but slipped sideways after flattening out. Note that it did not assume the often-mentioned teardrop shape.

Metamorphosis

A Stream of Water Becomes a Series of Drops...Joined Together by Narrow Necks of Liquid.

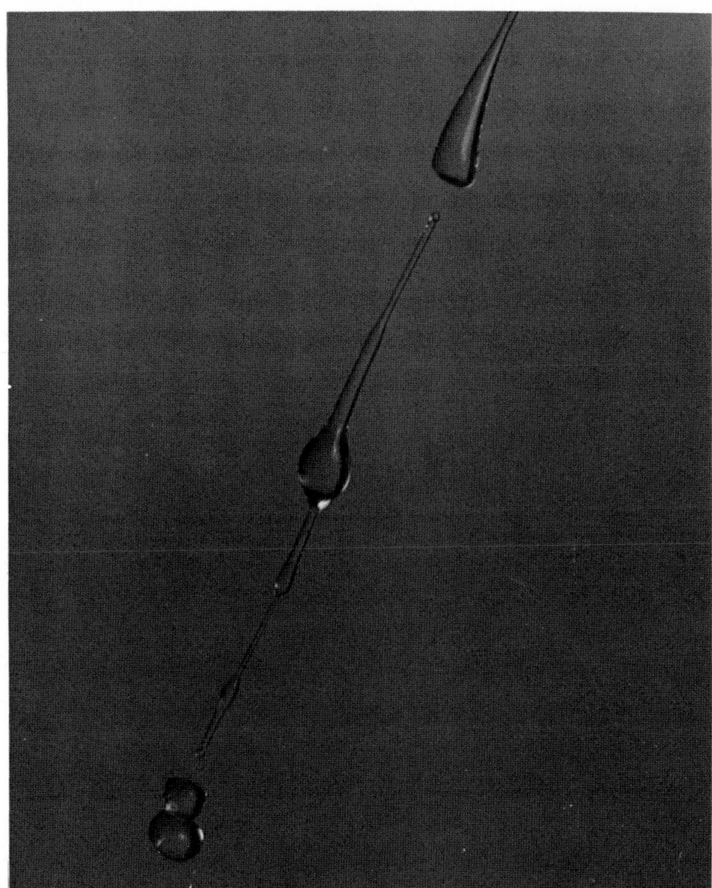

Water "Flowers"

The water "flowers" were created by a jet of water.
Charles E. Miller

Glassy Immobility

Plain tap water flowing from a faucet. Note that the water is streamlined near the faucet but quickly becomes turbulent. In 1/50,000 of a second even this turbulence poses with glassy immobility.

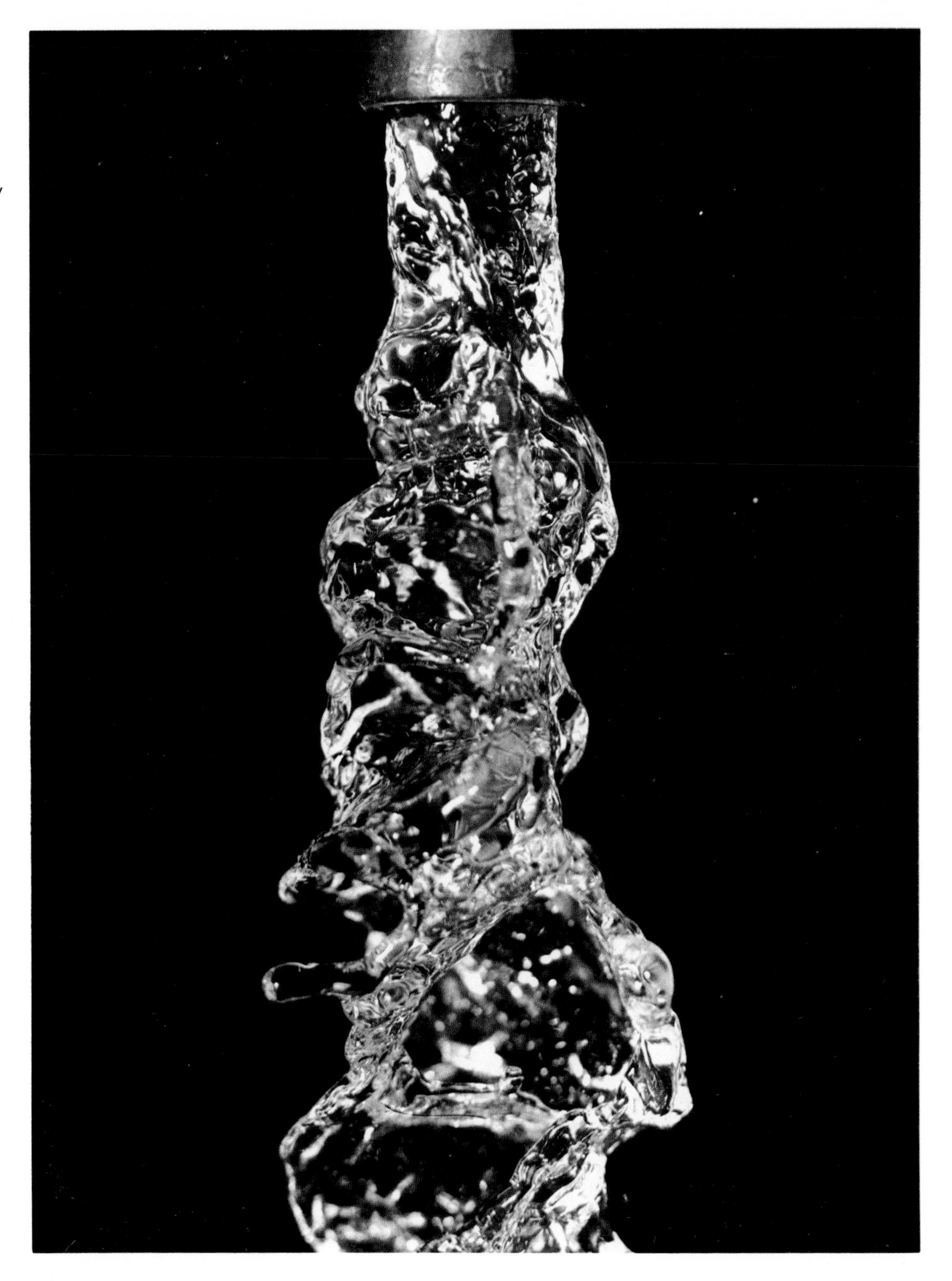

Portrait of Shock Waves

Jet of high-pressure air and adjacent, normally invisible, shock waves in the air. This is a striking example of how the stroboscope has achieved a new clarity in shadow or silhouette photography. See examples in the color section.

Arturo Rosales

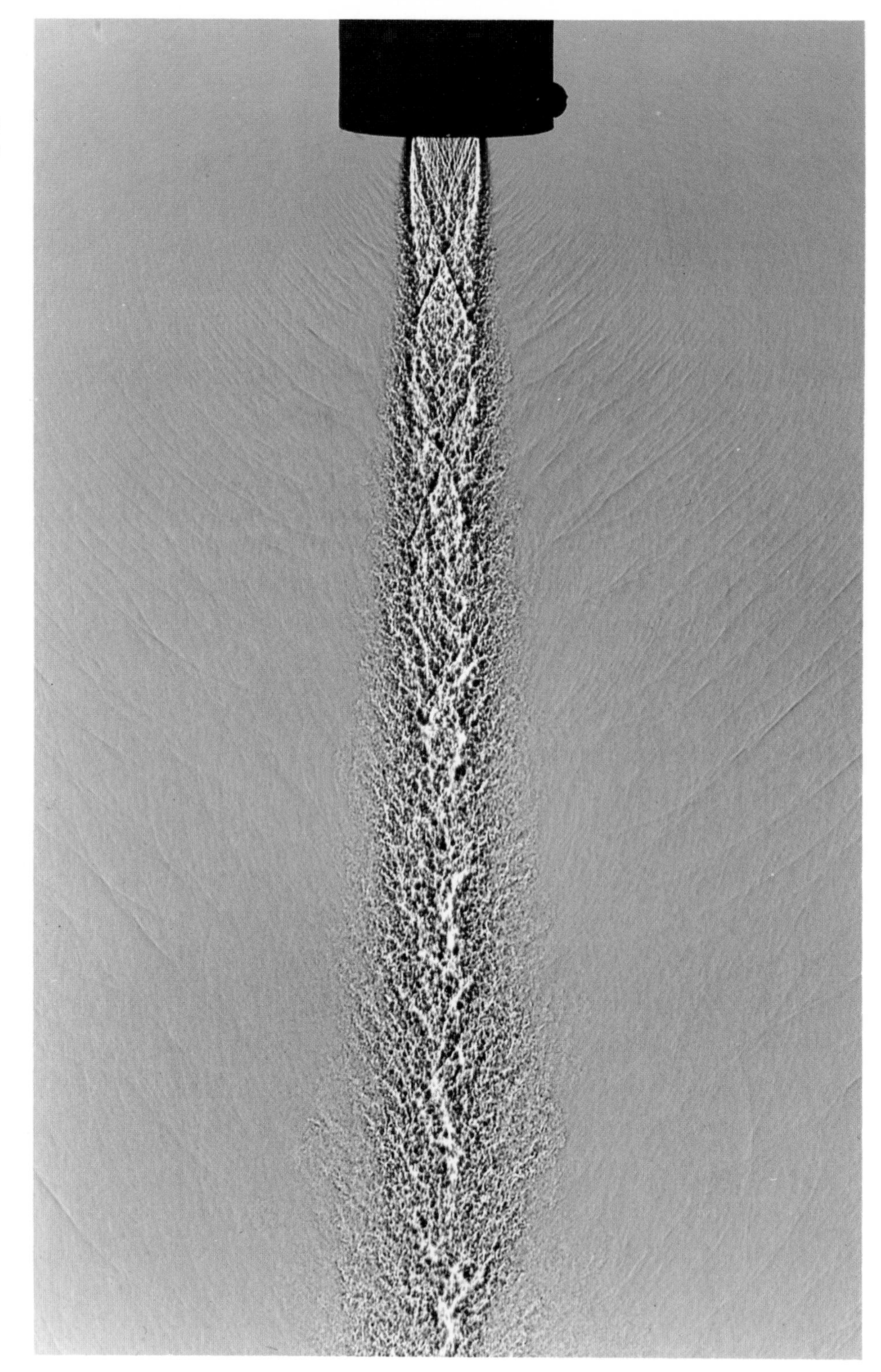

Bullets and Blasts

Many questions about the flight of projectiles remained unanswered until ways to photograph them were devised. Ordnance experts believed for years that bullets continued to gain speed for many feet after they left the muzzle of the gun by the continued pushing effect of the propelling charge in the open air. Silhouette photographs proved conclusively that this acceleration lasts for only a few inches (in the firing of a rifle, for example). Who has not heard it said that the kick of a pistol affects the accuracy of a shot? Yet photographs show that no uptilting of the gun occurs until the bullet is several feet on its way. Similarly definitive facts have been learned about the leakage of gas around the bullet in the barrel and about the speed of the sound created by the bullet.

Until the refinement of high-speed photography, bullets were usually studied only by shadow or silhouette photographs, a method used by the Austrian physicist Ernst Mach about 1881 and improved by English scientist Charles V. Boys. Others preceded Edgerton in achieving lucid shadow or silhouette photographs of bullets (1950), notably Hubert Schardin of the Scientific Institute at Weil. In shadow photography the object to be photographed is made to cast a shadow as it passes between an open-air spark and a photographic plate. This method is of great value because it will record sound waves and turbulence in the air, but it has several limitations. The photography sometimes must be carried out in darkness, the timing of the spark requires complex equipment, and the picture obtained, sharp as it may be, is only an outline. With other types of high-speed photography these limitations are avoided, and genuine reflected-light ballistic photographs are possible.

For examples of this achievement in photography and detailed views of the spectacular events that occur when guns fire and bullets strike, turn the page.

The Splash of a Bullet in Three Acts

When a .22-caliber bullet strikes a steel block, it apparently liquefies from the force of the impact and splashes in a manner not unlike the milk drops. As the liquid lead splatters and solidifies again, the particles radiate outward in the charming concentric-circle formation seen in the adjacent picture. The origin of the concentric circles is probably grooved marking on the bullet. Highlights in the pictures of the splatter show that the liquid lead has a peculiar surface like molten or polished silver. Exposure is 1/1,000,000 of a second.

Inspection of an armor plate after being struck by high-powered projectiles shows that the bullet momentarily acts as a liquid.

Two Stages in the Firing of an Old Revolver

The trigger is pulled, and the action starts. The bullet has not reached the muzzle, but gas is ejected. In ordinary light the gas is hardly visible. Exposure is 1/1,000,000 of a second at f11.

The bullet is out but is still surrounded by gas. The propelling charge rushes out of the muzzle, and powder particles from a previous shot are driven ahead of the gas.

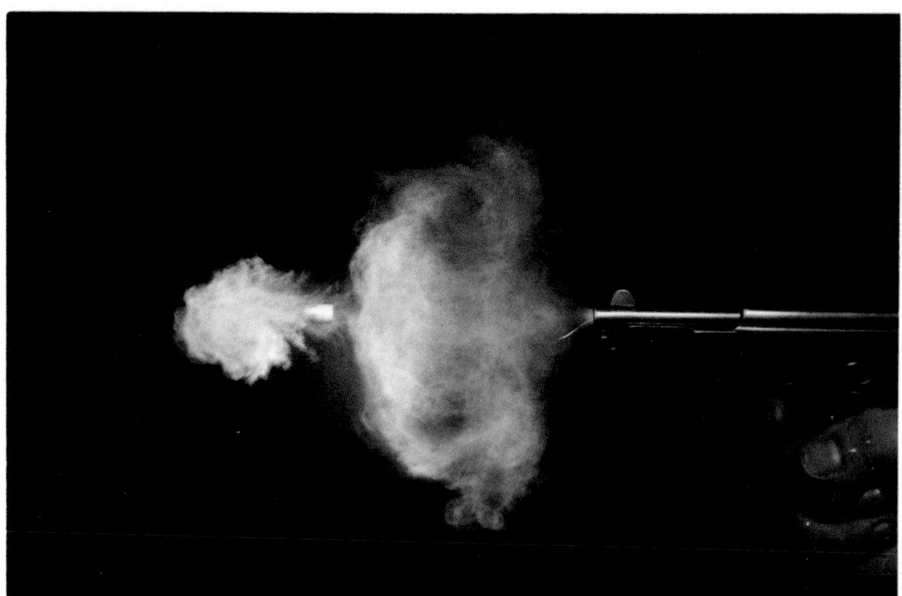

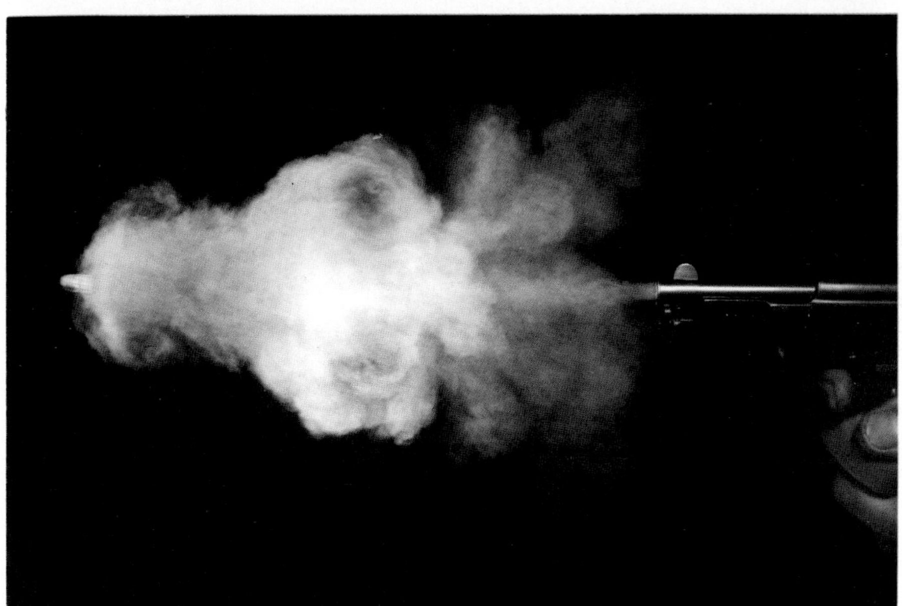

Firing a Mauser Automatic

Contrast this sequence with that of the 1878 revolver on the preceding page. In picture 1, the bullet is just outside the barrel but still behind the vanguard of leaking gas; in 2, it has drawn ahead of its propelling charge and is entering the gas; and in 3, it is moving beyond the gas and creating vortices.

By using the sight of the gun as a reference, stages of the return of the barrel from its recoil may be noted. The round object in the foreground is a microphone which starts the stroboscopic light when it picks up the sound of the discharge. The distance of the microphone from the gun governs the timing of the photograph since sound travels at a known velocity (about 1,100 feet per second). Exposure is 1/1,000,000 of a second.

Death of a Light Bulb

The speed of this .30-caliber bullet is approximately 2,700 feet per second, or over 1,800 miles per hour. Cruising along at this speed, a bullet encounters an electric light bulb, and the high-speed camera reports the details of the collision. The action shown in the four pictures lasts about 3/10,000 of a second, each picture being exposed for one-millionth of a second.

In picture 1, a double-exposure view shows the bulb before and after impact; in 2, even the crimp marks where the shell was attached to the bullet are clearly visible, as are the cracks traveling faster than the bullet; in 3, as the bullet enters the bulb, the compression wave (traveling about 15,000 feet per second) in the glass cracks the other side before the bullet gets there; and in 4, the bullet is several inches away from the bulb and is emerging from the arrowhead formation of gas and glass dust that accompanied it out of the bulb. A split second later the bulb becomes a cascade of falling glass fragments.

The four pictures were not taken in succession; each is of a different bullet and a different bulb caught at different phases of the action.

Bullet Clips a String

A cylindrical bullet moving at 4,000 feet per second from left to right is shown by the microsecond exposure just after the bullet has clipped a string. Apparently a short piece of the string remains after the bullet has passed.

Further photographic studies reveal that a string or other object emits light at the instant of high-speed impact, thus giving a brief self-exposure.

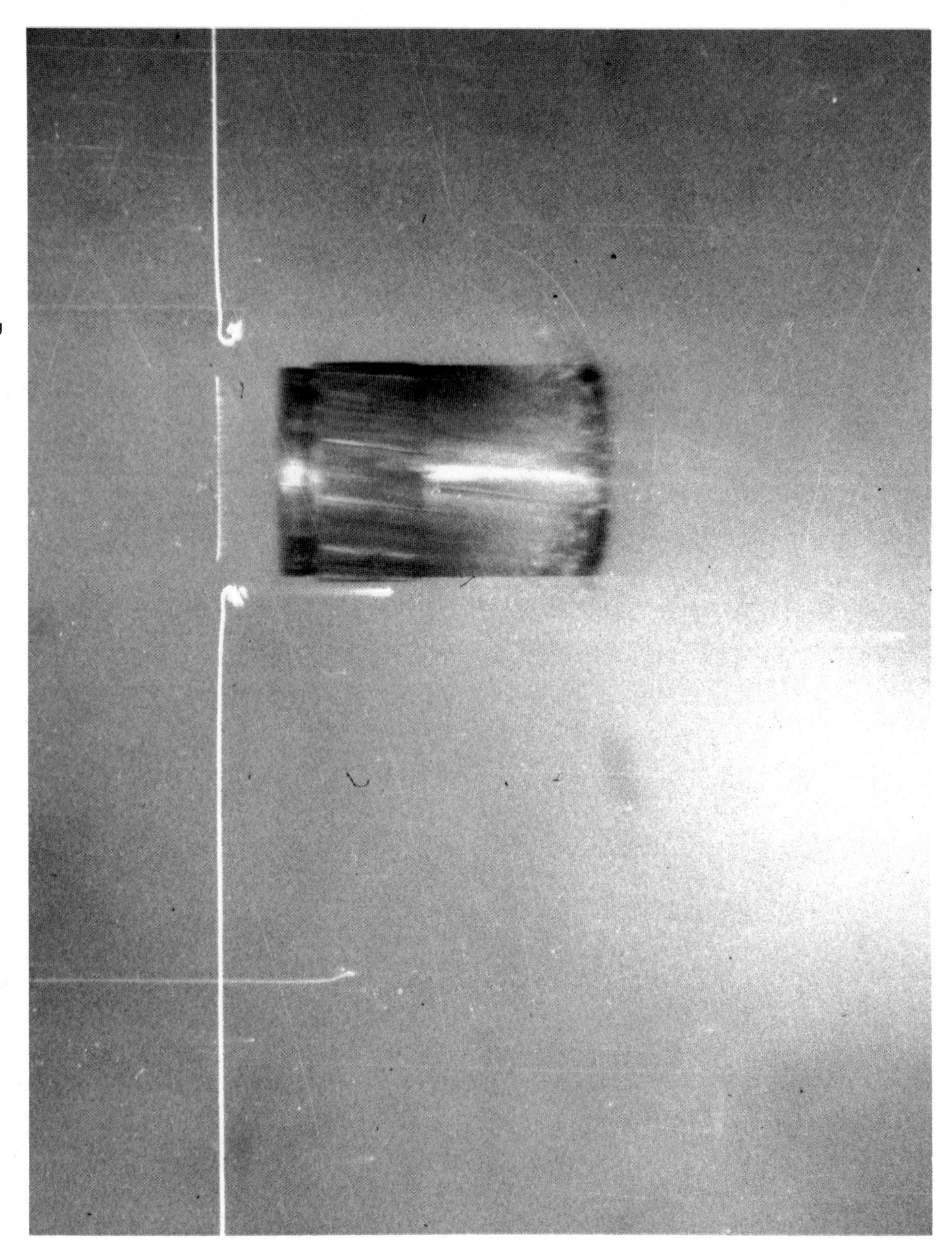

Bullet Plucks a Copper Wire

Multiple exposure of a .22-caliber bullet cutting a single copper wire. (1962)

Harold E. Edgerton
Courtesy of the Polaroid Corporation

Bullet Smashes Plexiglas

Silhouette photograph of a .30-caliber bullet traveling at approximately 2,800 feet per second. Photographed with an exposure of less than one millionth of a second immediately after it penetrated a strip of Plexiglas. (ca. 1965)

Usually in photographs of this kind the flash is initiated by the shock wave. Note, however, that in this picture the shock wave has not yet reached the microphone (the square black object at the bottom of the picture). The flash was initiated instead by a faint shock wave that traveled down the Plexiglas from the point of impact and then out to the microphone. Sound travels ten times faster in Plexiglas than it does in air.

The white wispy material beside the Plexiglas was caused by light reflected from the surface of the Plexiglas onto the film. The dark spots at the right were caused by pieces of Plexiglas splattering on the film.

Shock Waves Revealed

At the right, a .22-caliber "long" rifle bullet traveling faster than the sound in air is photographed in silhouette to show its shock waves. Note that waves have a dark and a light side. At the left, a .22-caliber "short" with a velocity less than sound traveling in air is shown with its sound waves. Note that the waves at the back appear as light lines with no dark phase, and that some are almost straight.

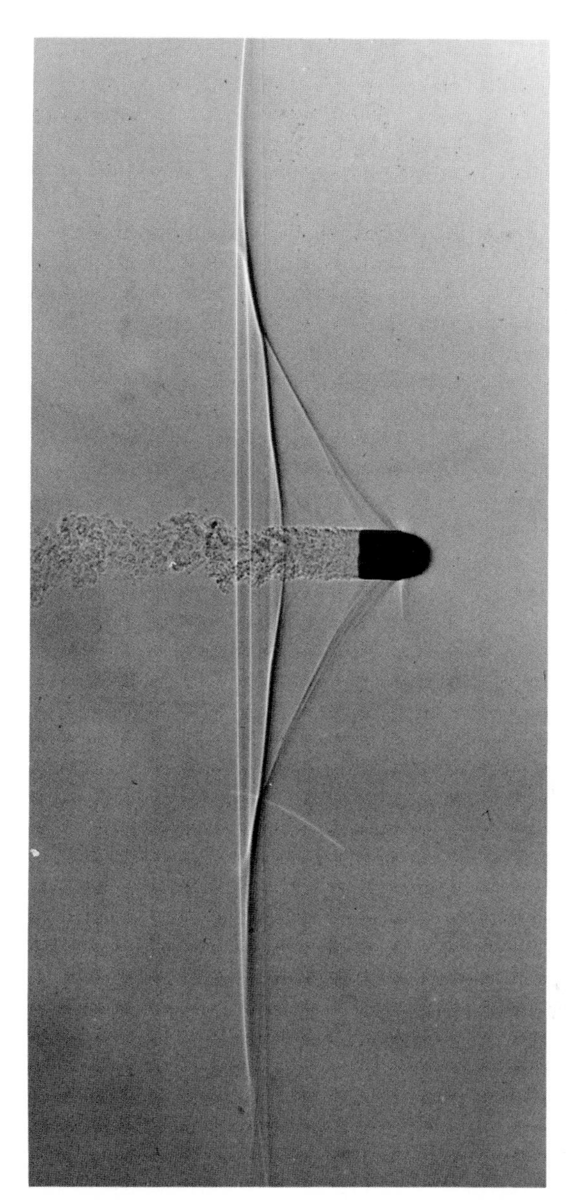

Penetrating the Bubble

A .22-caliber rifle bullet penetrating a soap bubble filled with argon gas. Note the shock waves — thin lines that one does not often see.

High-Speed Particles

High-speed particles ejected from the end of a dynamite cap. Velocity of the particles is about ten times the velocity of sound. The exposure is about 1/100,000,000 of a second.

Exploding Dynamite Cap

An exploding dynamite cap and the shock waves produced by its fragments. The dynamite cap was in a steel tube to protect a field lens 10 cm. in diameter and a spark light source.

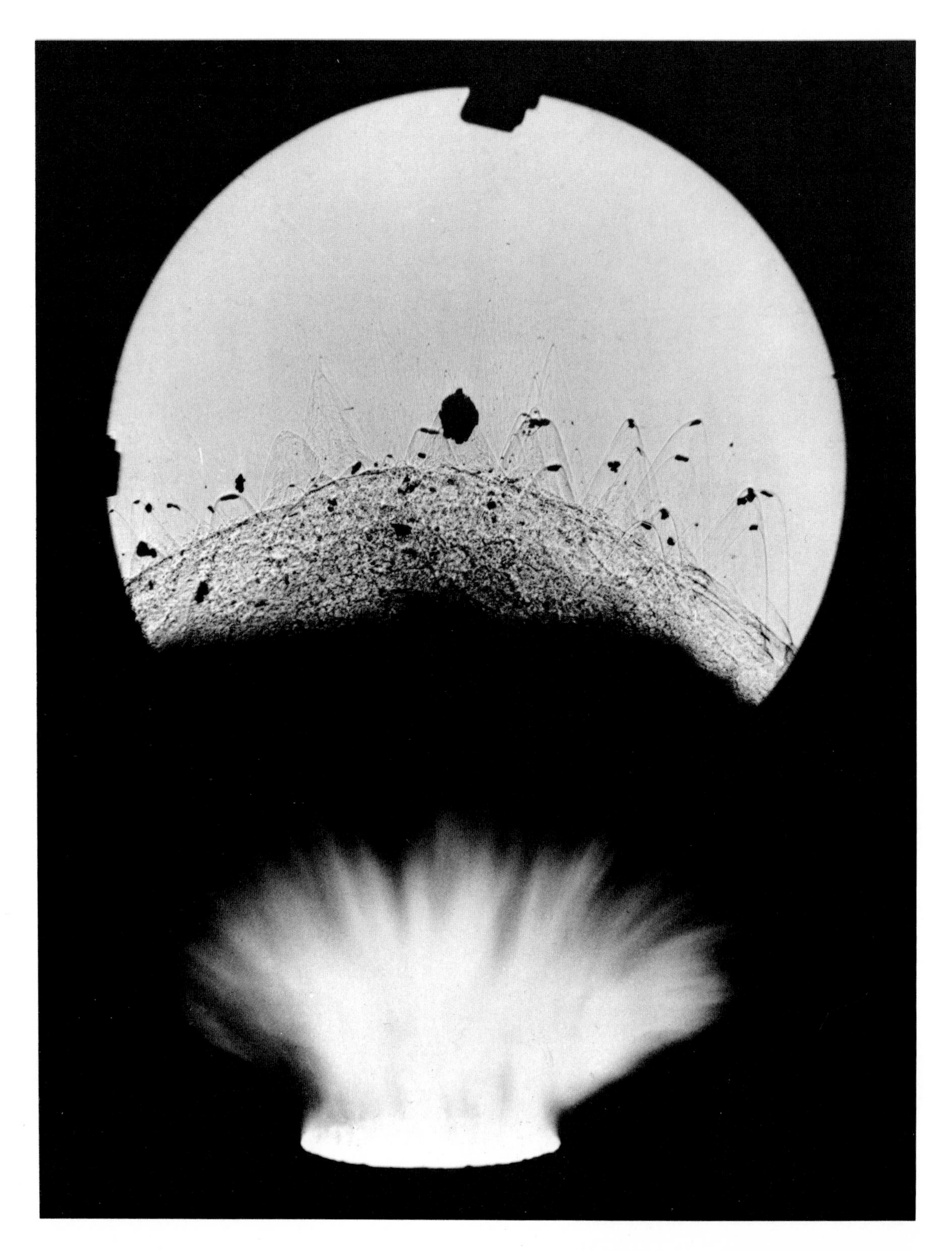

Exploding Pentolite

One microsecond exposure of a 1-inch-square stick of explosive Pentolite. The detonation was initiated by a dynamite cap at the far end of the stick. The photograph shows the shock waves coming from the flat sides of the square material and not from the corners as expected. The photograph was taken just before the detonation wave reached the end closest to the camera.

In the second picture, a side view, the magneto-optic shutter opened several times to produce a multiple exposure of the incandescent shock waves produced by the explosion of the Pentolite. Notice that the shock waves off the end are very strong.

The bottom part of the second picture showing the tandem pieces of Pentolite was, of course, taken separately from the one above showing the explosion.

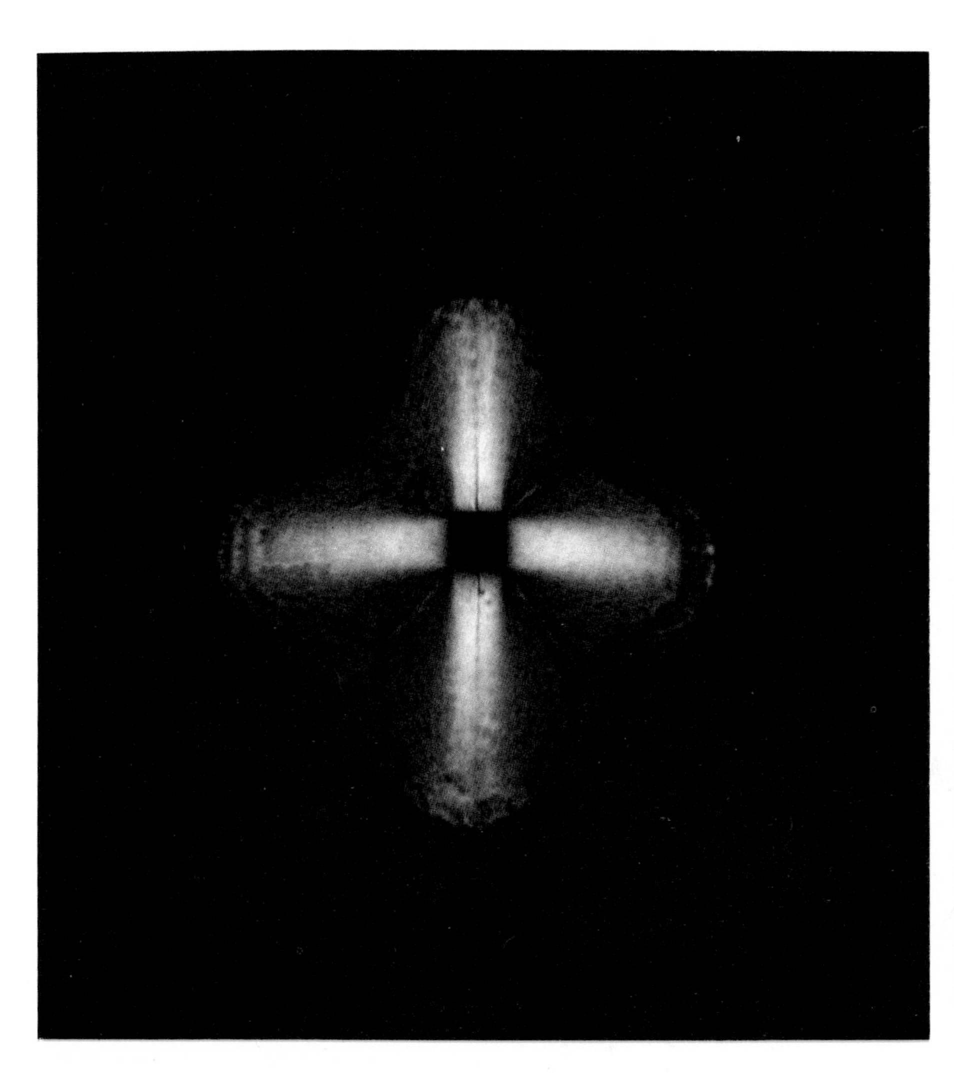

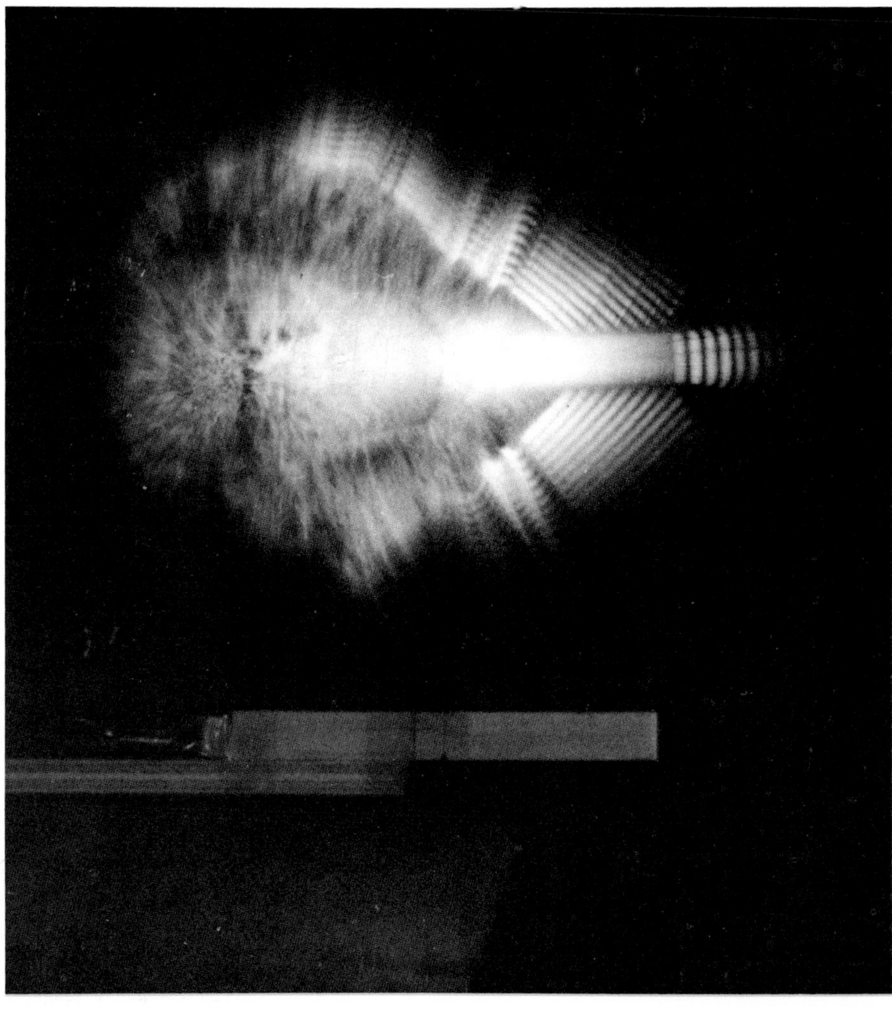

Jack of Hearts

The Jack of Hearts is cut by a .30-caliber bullet. Exposure is less than a millionth of a second. (1960)

Shadow Photography

Barry Rosoff fires a blank charge in a revolver to illustrate the use of a Scotchlite screen to obtain shadow photographs of a large subject in normal light.

A small strobe lamp is placed very close to the camera lens which is focused on the Scotchlite reflecting screen. A shadow of the gun and the shock waves is produced by the small lamp.

The Scotchlite screen is covered with small glass spheres that reflect most of the light directly back to the strobe lamp and lens.

The camera shutter is open a short time while the phenomena are initiated.

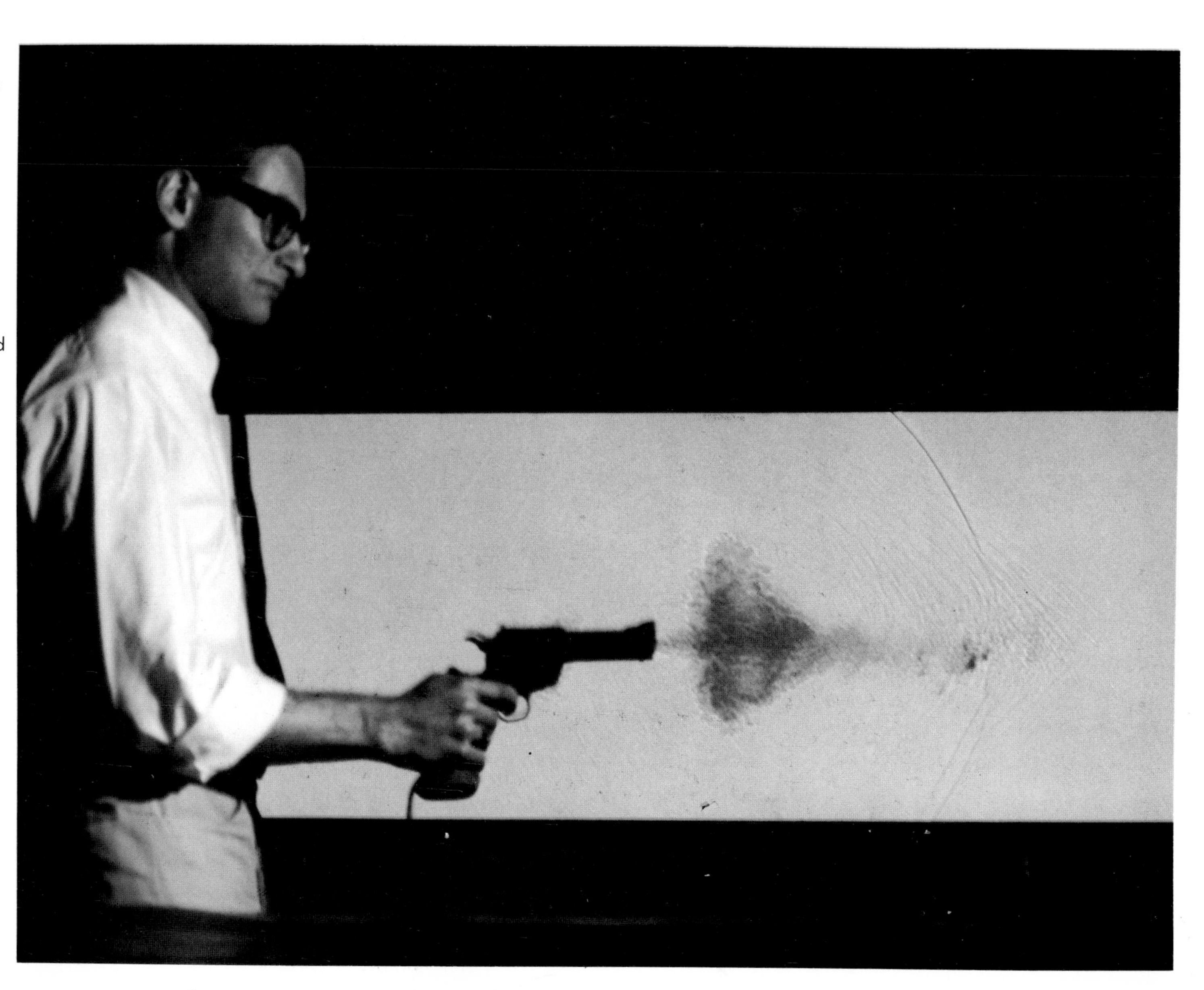

Living Motion

Birds and Beasts and Insects Dire

Does a trotting horse ever have all four feet off the ground at once? In 1872 this was an agitating question to a group of California sportsmen, for bets had been laid and noses were out of joint over the contradictory answers provided by the unaided eye. When a definitive answer was obtained, not only were bets settled but a long step forward had been taken toward the achievement of motion pictures and the photographic recording of motion.

The familiar story is worth retelling. It was Leland Stanford, railroad magnate and former governor, who hit upon the idea that the fledgling art of photography might solve the problem. He sought a photographer and found Eadweard Muybridge, then plying his trade in California. Muybridge enthusiastically went about photographing a trotting horse and, with a few tantalizing exceptions, consistently missed the horse or drew blank negatives because his method of timing was inadequate and fast emulsions were not then available. Undaunted, Stanford called in one of his gifted engineers, John D. Isaacs, to devise a system of timing, which Isaacs promptly did. With a battery of cameras lined up along a track and with special shutters operated either electrically or by a clock, he made it possible for Muybridge to take a whole series of properly timed instantaneous pictures of the trotting horse. The result was sensational; not only did the pictures show the horse at times with all four feet off the ground but they revealed facts about the locomotion of the horse that left even horse fanciers surprised. Muybridge's career was made, and he spent the rest of his days photographing all manner of creatures including elephants, asses, hogs, deer, men, baboons, and kangaroos using this method.

Stanford proudly showed his pictures in Europe and sent Muybridge over with the others in 1881. The pictures not only served to settle an argument among French artists over the attitudes of a running horse but it was quickly recognized that they could be used in the projecting zoetrope, a device based on Plateau's stroboscope to synthesize a series of pictures into an illusion of motion, of a horse actually trotting. More important, the pictures came to the attention of Dr. E. J. Marey, physicist and founder of the Marey Institute, who had spent years studying animal motions. He immediately spotted the defect in the Muybridge pictures: when they were viewed in the zoetrope, the horse stayed in the same place and the scenery ran by. He then proceeded to build a photographic "gun," based on a camera built by the astronomer Janssen to record the transit of Venus, that would take a series of pictures from the same point. After the development of this gun, the motion picture as we know it was just around the corner, awaiting flexible film.

The pictures taken by Muybridge in America and by Marey in France proved dramatically useful in photographing live subjects and analyzing their motions. Artists used them as study material; physiologists, anatomists, and naturalists gained new knowledge from them; and laymen for the first time penetrated the fascinating world that lay beyond the blur of rapid motion.

Against this background it remained for Professor Edgerton to apply his stroboscopic system to perfect the photography of ultra-rapid motion of living creatures. On the following pages are examples of his pioneering pictures of birds and bats in flight and of dogs and dolphins in action.

Once Edgerton had invented the basic technology of photographing the ultra-rapid motion of living creatures in flight, others used his system, or refinements of it, to achieve portfolios of photographs of thrilling beauty and scientific value. Notable examples, by both amateurs and professionals who have achieved superb photographs of birds, are included in the colored signature of this book. For example, there are remarkable hummingbird pictures taken by Crawford Greenewalt, sometime president of the DuPont Company, and the work of Don Bleitz.

Hummingbirds

At Holderness, New Hampshire, the late Mrs. Laurence J. Webster for many years made her home a sanctuary for ruby-throated hummingbirds. These exquisite creatures, the smallest of American birds, weigh only about one-tenth of an ounce and beat their wings so rapidly in flight that they appear only as a blur. Mrs. Webster attracted them not only by providing the flowers they love but by placing brightly colored vials containing sweetened water about her yard. They became so accustomed to her hospitality that they alighted on her outstretched finger.

One day Vannevar Bush, then professor of electrical engineering at MIT, challenged Edgerton to take a photograph of a hummingbird that did not show its wings as a blur. Efforts by photographers to record the flight of the hummingbird by conventional cameras and lights had always shown the wings as only a blur. Edgerton, for the moment, was stymied. "I will photograph a hummingbird," Edgerton said to Bush, "if you can catch one to pose before my camera." Some time later, Bush, having heard about Mrs. Webster, gleefully told Edgerton he was calling his bluff. So Edgerton promptly carried his early high-speed equipment to Mrs. Webster's home and, with her help, succeeded in taking both stills and motion pictures, probably the first photographs ever taken that showed the wings without a blur. The results are shown on the following pages. Exposures of 1/100,000 of a second brought the motions of the birds to a dead stop, and high-speed motion pictures taken at 1,000 per second made it possible to measure the frequency of the wing beats; it was found that the wings beat about 60 times a second when the birds are hovering and about 70 a second when the bird is frightened and flies away.

Edgerton's achievement inspired others, notably Crawford Greenewalt, to study the flight of both hummingbirds and other birds. In the beginning of his hummingbird photography, Greenewalt was successful in getting colored pictures of hummingbirds standing still, but in flight their wings were always blurred because he was using standard amateur or professional commercial flash units. He next tried faster commercial microflash units with exposures of 20 microseconds. These units gave beautiful black and white pictures in which the wings were sharp and clear, but the lamp was too weak to achieve good color.

Greenewalt was a member of the MIT Corporation and while in Cambridge he visited Edgerton in his laboratory. With his accustomed generosity, Edgerton gathered about a team of students to produce color without blur. He had already constructed equipment at the request of the National Geographic Society and was aware of the work of Arthur A. Allen of Cornell, the great ornithologist who had earlier used electronic flash for bird photographs. This type of flash unit did not meet Greenewalt's specifications, so Edgerton worked on other designs that were light and portable. Greenewalt had turned to electronic experts in Wilmington for their assistance, but it was Edgerton who set him on the right path. Greenewalt's adaptation of the strobe light with automatic triggering systems resulted in an outstanding set of photographs. To express his appreciation, Crawford Greenewalt presented Edgerton's laboratory with funds to purchase a cathode ray oscilloscope, equipment that has been very useful to Edgerton for observing the output of flash lamps.

Greenewalt photographed his first hummingbird in 1953 and went on to take spectacular photographs of hummingbirds from many locations in North and South America. He also studied the flight of birds, the irridescence of hummingbird feathers, and the methods by which birds produce their songs. Altogether an impressively unusual achievement for a chemical engineer and a distinguished industrialist!

I report the work of Greenewalt as an illustration of how Edgerton's pioneer work provided the methods and inspiration for others to put his basic discoveries to work in unexpected ways.

The late Mrs. Laurence J. Webster, who befriended hummingbirds, was photographed in 1936 by Edgerton.

Probably the first successful photograph ever made that showed the wings unblurred. Note that the bird is sticking out its slender tongue. Edgerton also took movies and colored photographs of these birds.

Fighting hummingbirds contending for a place at the feeder.

With a grant from the National Geographic Society, Edgerton, together with the late R. J. Niedrach and Walker Van Riper, photographed hummingbirds in Colorado in 1947. Edgerton then went on to California where the Anna's hummer *(Calypte anna)* was photographed at the Dorothy May Tucker Sanctuary in Santiago Canyon.

The pictures taken in Colorado provided an explanation for the strange "trilling" or vibratory sound made by broad-tail hummingbirds. The source of this "shrill trilling" is the vibration of the tip of the primary and occurs not when the birds are in normal flight but when they approach and challenge each other.

Black-chinned hummingbird photographed in California in 1947. The bird's wings make a buzzing noise that sounds like a bumblebee.

Flight of a Dove

New England Bat

The first two pictures are of the little brown bat *(Myotis lucifugus lucifugus)* common in New England. While a Harvard student, Donald R. Griffin, who had been studying these tiny "flying mice," brought specimens to Edgerton to be photographed.

Stalling to avoid a crash into an adjacent wall

The membrane that joins the tail and hind legs of this bat serves as a horizontal rudder.

Trinidad Fishing Bat

The *Noctilio leporhinus* seeks its food on the surface of water.
Bat provided by Donald R. Griffin

Revealing the Secret of the Mexican Freetailed Bat

Unlike the bats in the preceding pictures, the thousands of bats *(Tadarida brasiliensis mexicana)* that make their home in the Carlsbad Caverns in New Mexico appear to have no rear membrane to serve as a control surface. This had long puzzled zoologists and bat fanciers, but Edgerton solved the puzzle by revealing that these bats have a retractable membrane behind their hind legs. He went to the caverns and with arduous labor set up equipment to photograph the bats as they flew in and out of the cave. These pictures showed that a membrane unfurls from the long, thin tail when the bat is flying and is retracted to conserve body heat and moisture when the bat is at rest. The extended membrane also serves as a scoop for food.

Happy Dog

The acknowledging end of a friendly dog

Mosquito Bite

The *Anopheles* mosquito biting. Its stinger is covered by a sheath or labium which is split so that when the mosquito inserts the stinger (fascicle) it is lubricated by liquid but only the stinger penetrates the skin of the victim. In the June 1978 issue of *Scientific American*, Jack Colvard Jones of the University of Maryland wrote an extensive description of the "exquisitely made piece of biological machinery" by which another kind of mosquito, the female *Aëdes aegypti*, bites.

When the mosquito acquires its blood sample, it strains out and digests red corpuscles and ejects liquid plasma. Practically all photographs, especially color, taken of small, live subjects are made with electronic flash lighting.

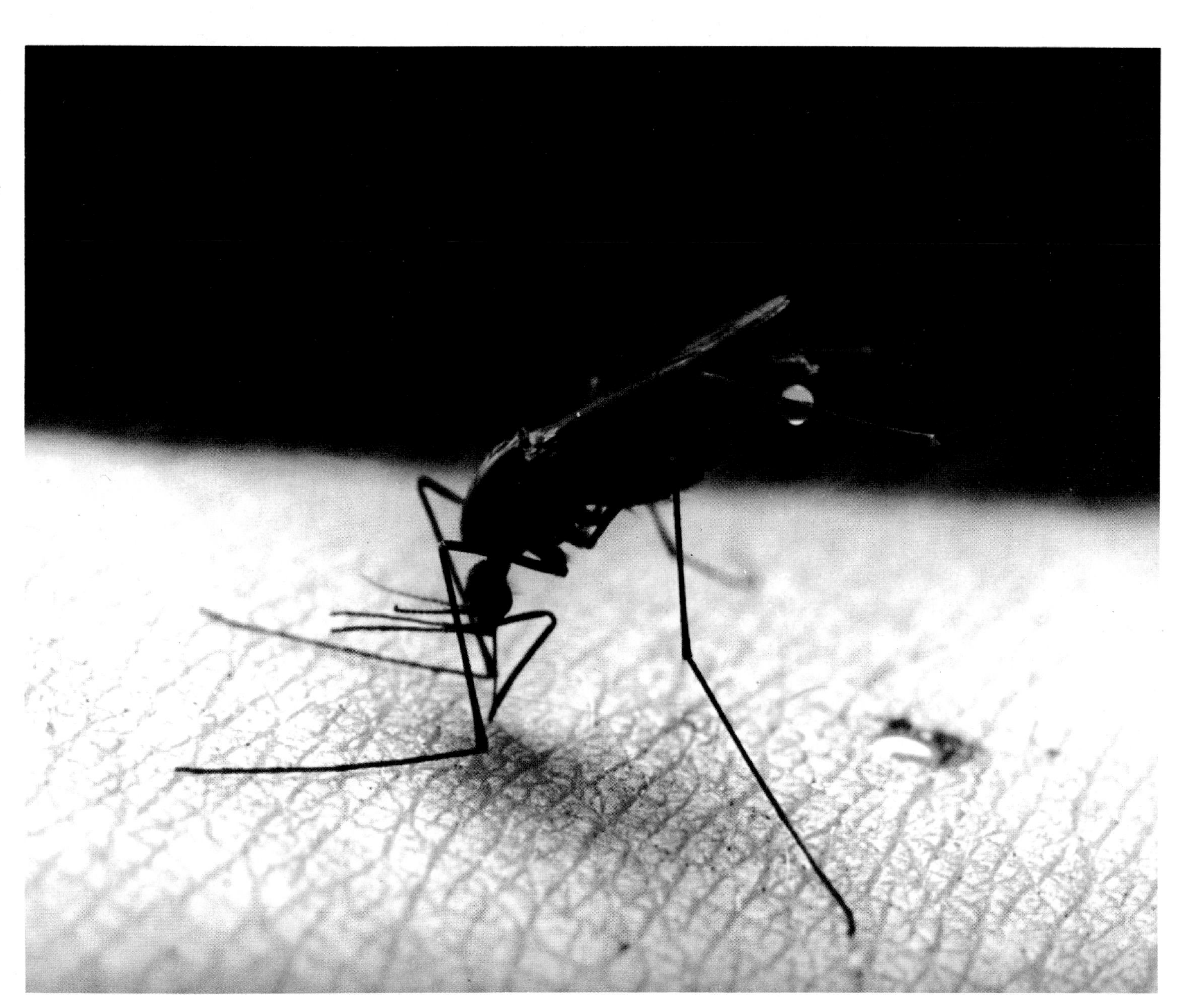

Penguin

Penguin diving in an aquarium tank

Dolphin

A performing dolphin at the New England Aquarium

Show-Off

The dolphin swings it, en rapport with the audience at the New England Aquarium.

Sports

By recording motion in relation to time, the motion picture has contributed to almost every sport in which speed and control are important. Its chief deficiency has been its slowness; the speed range of the most modern cameras designed for motion study is too restricted to provide a useful record of many actions.

Ultra-high-speed photography, both motion and still, has broken through this restriction and provided a means for studying form and technique equal to any speed within the capacity of performers. It has been in the field of sports that stroboscopic multiple-exposure photography has been most extensively used. Examples of the application of high-speed still photography to golf, tennis, football, and other games are shown in the pages immediately following.

When the Ball Is Struck

Here are four studies of what occurs when a golf club strikes a golf ball. In the last picture note the egg-shaped form assumed by the ball. For a period after its compression by the club the ball oscillates, swelling and shrinking along the horizontal axis.

A Drive

Multiple-flash (300 per second) photograph of a driver, showing the distortion wave running up the shaft after the ball is hit. The stage of the stroke showing the s-bend in the shaft comes a little less than 1/1,000 of a second after contact. The image of the traveling ball between alternate images of the club shows that the ball is traveling twice as fast as the club head after impact.

Iron Shot

Single flash of the late Bobby Jones making an iron shot. The squashed ball indicates that the strobe light flashed at the exact moment of impact. Careful observation of golf shots indicates that the head leads the shaft at the moment of impact.

Catching the Click

Is there a golfer who has not wondered what happens during the "click"—that all-important event when the club imparts its energy and the intentions of the player to the ball? And how many inconclusive arguments have been waged in locker rooms over the questions:

1. Is the follow-through really important?
2. Does overspin of the ball rather than backspin ever occur in normal play?
3. Does the club impart the spin, or is the spin produced by the action of the air on the ball?
4. Does a heavy club produce a longer drive than a light one?

The pictures on the following pages, taken by single-flash and multiple-flash techniques, show the details of the click.

When these photographs are analyzed, we see clearly that the club and ball are in contact for a brief time and that they travel together for only a very short distance. The time of contact is about six thousandths of a second, and the distance they travel together is much less than an inch (about 0.3 inches). From a golfer's standpoint, the ball is off the club face when the face has advanced to a position over the center line of the tee. During this extremely short interval the ball is compressed and springs away from the club head. Thus the club imparts its momentum to the ball by impact, and there is no follow-through action whereby the club continues to push and direct the ball. It is impossible for the golfer to manipulate the ball while it is on the face of the club.

A black cross is painted on the ball in order to determine the spin. Measurement of the angular shift of these lines between flashes, combined with the known time between flashes, makes it possible to determine the speed of rotation. All normal shots show backspin, from about 2,000 revolutions per minute for balls hit with drivers up to approximately 10,000 revolutions per minute for balls hit with more lofted clubs, such as a No. 7 iron. Since a ball is in the air for about six seconds, it turns over 200 times in traveling from tee to turf, assuming constant speed of rotation.

The velocity of a golf club increases as the swing progresses, as shown clearly by the multiple-flash photographs taken at uniform intervals. Just before impact, the velocity may decrease slightly, as indicated by the forward bending of the club shaft. Careful examination of the photographs shows that the head leads the shaft just before impact, meaning that the club head is actually being slowed down by the shaft.

Bobby Jones would not believe that this was true of his strokes. He thought it was a camera illusion, such as the highly distorted photographs taken by a focal-plane shutter camera. Edgerton demonstrated to Jones that a strobe-illuminated photograph records the entire scene at the instant of flash so that only the golfer is distorted.

When they see multiple-flash photographs of strokes, expert golfers are usually surprised to discover that the ball travels faster than the club. If the ball and club were perfectly elastic, the ball would travel nearly twice as fast as the club in accordance with the principle of the conservation of energy and momentum. Practical balls, however, are not perfectly elastic, and experimentally their velocity is found to be only about 40 percent greater than that of the club. A heavy club (head weight 11.6 ounces, for example) gives an increase in speed of about 47 percent, while a light club (head weight about 7.3 ounces) gives a 36 percent increase. The light club, however, is usually swung faster, making up for the difference.

Densmore Shute Bending the Shaft

Golfers had best interpret this swing analysis for themselves, but laymen will notice the bend in the shaft after the ball is struck. Interval of time between pictures is 1/100 of second. (1938)

James Thomson Bending the Club

In this drive, the marked bending of the shaft can be seen, as well as the way the golfer wraps the club around his neck at the beginning and end of the stroke. Notice the tee flying off to the right after impact.

Ralph Guldahl Bending Shaft on a Drive

In this photograph, a straight edge along any one of the positions of the club's shaft shows that it bends on a drive.

Toe Shot

Here is an ordinary golfer. It is a toe shot, and the twist of the club head is readily seen.

Compare this stroke with those of Jones, Guldahl, Shute, Thomson, and other top-flight golfers on the preceding and following pages.

Dub Stroke

The multiple-flash strobe records a dub stroke as the club strikes the ground as well as the ball at 120 flashes per second.

Impact of an Iron — 1,000 Pictures per Second

The ball turns over in four pictures, which corresponds to 4/1,000 of a second or a rotational speed of 250 revolutions per second. This is a difficult measurement to make without multiple-flash photography.

Tumbling Tee

Tennis Impact

In the pictures of tennis players on the following pages further evidence is found of the analytical power of single-flash and multiple-flash exposure photography. Both techniques are being used for the study of sports equipment, and both may be used to investigate the principles of impact and flight as they affect performance. In the single-flash picture note the squash of a tennis ball and the indention of the strings as the ball is hit. These pioneering tennis photographs were taken in the 1930's.

72–73

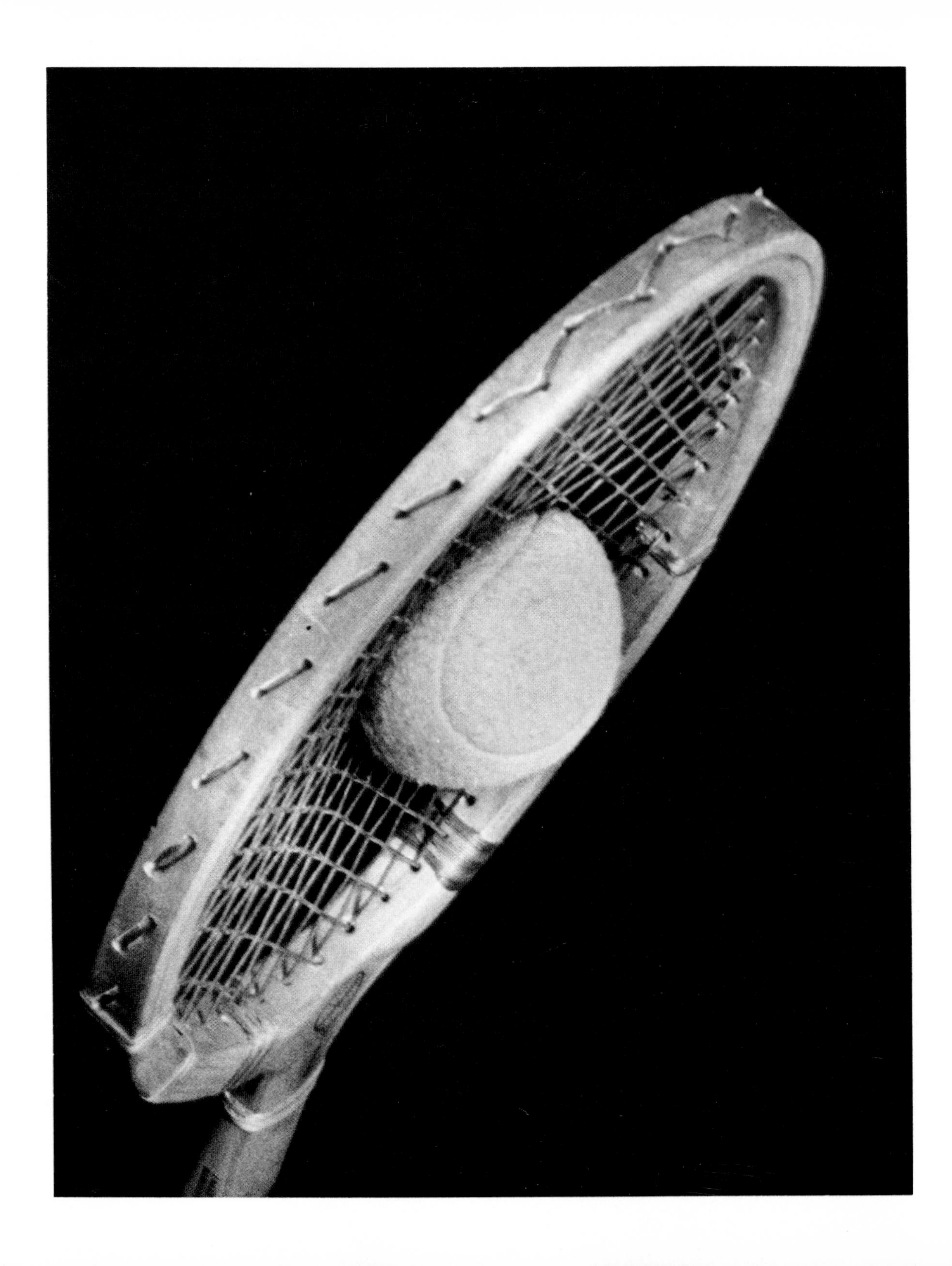

A Serve

Here the flashes were timed in relation to the ball and racket so that the moment of impact was caught exactly. Note that the ball is flattened on both back and front sides. Analysis shows that the tip of the racket vibrates back and forth after the impact. John Bromwich is the server.

300 Pictures a Second

Multiple-exposure close-up of impact in a serve. The strobo-scopic light missed a flash just before impact, but the distortion of the racket, backward and forward, is clearly seen.

Swirls and Eddies

Multiple-flash photography, by coalescing successive images, captures the entire flowing record of a stroke. Note the trailing blur of the player's profile. (1939)

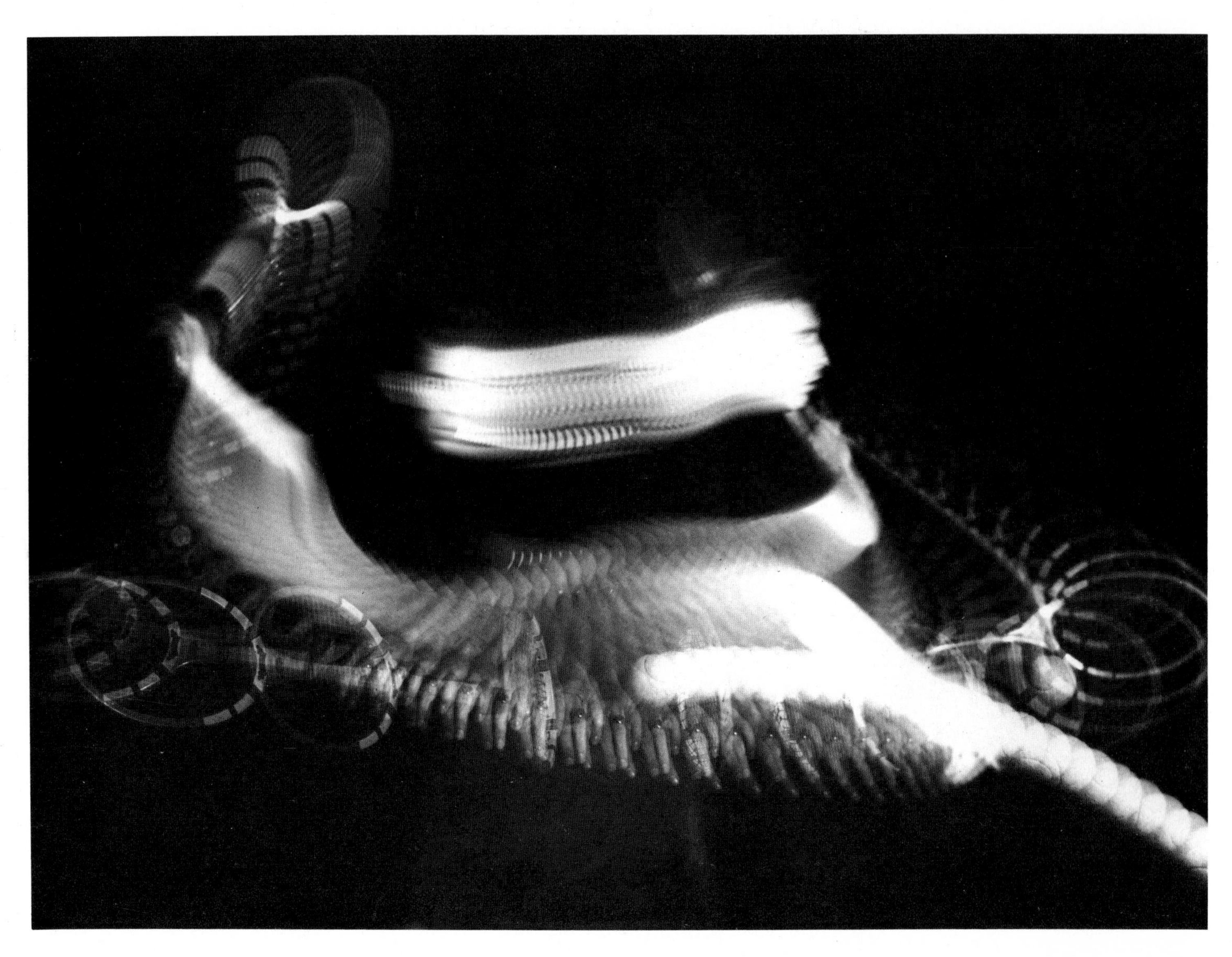

Gussie Moran (1952)

Placement Kicks

Every football fan has seen a placement kick, but no one has observed what actually happens in the fraction of a second when the booter's toe meets the pigskin. In the first picture, note how the toe of the shoe is pushed backward. In the second picture former Harvard coach Wesley E. Fesler, onetime all-American star at Ohio State, kicks a ball inflated to the normal playing pressure of approximately 13 pounds to the square inch. Measurements show that the boot penetrates at least half the diameter of the ball. After leaving the foot, the ball pulsates, swelling and contracting as it recovers its usual contour. (1934)

Note the fan of suspended dust that the top of the ball leaves behind. This is another classic in the history of photography.

A Simple Salute

Joe Levis, five times national foils champion, executes a salute that a fencer gives before he engages his opponent. In the next two photographs Levis instructs a student. An MIT alumnus, Class of 1926, Levis still finds time to inspire MIT students in the art of fencing.

Bent Bat Bumps Ball

This picture shows the bent bat as it contorts the ball.

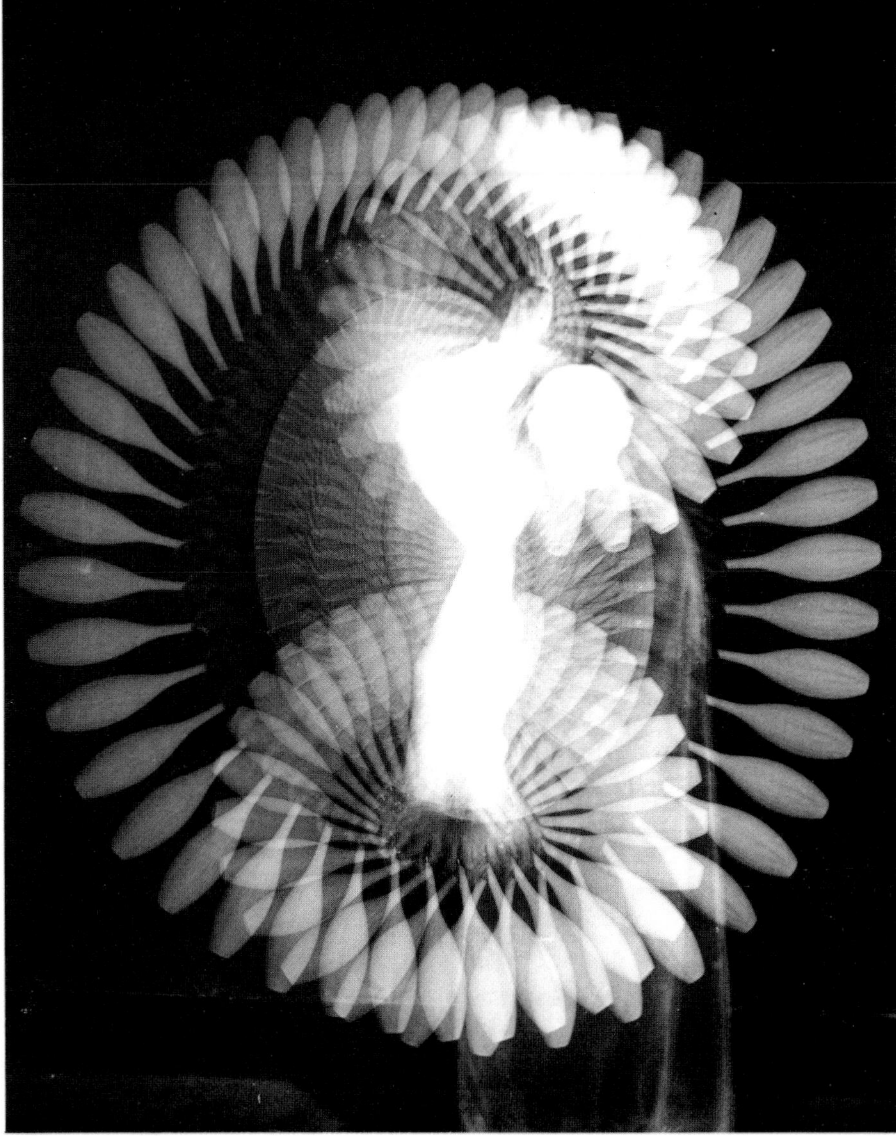

Indian Club Routine

The first of these two photographs is a result that can be obtained with a regular camera and is similar to the technique of recording abstract light patterns with which Ladislaus Moholy-Nagy of the Bauhaus experimented. The second of these pictures shows the same routine recorded by stroboscopic light.

Joe Louis vs. Arturo Godoy

The ferocious expression of heavyweight champion Joe Louis during his fight with Chilean Arturo Godoy in 1940. To prevent further injury to Godoy, the referee ended the bout in the eighth round, giving Louis his eleventh win.

Joe Costa, *New York News*

Basketball Game

Gjon Mili

Pole Vaulting

Once pole vaulters used stiff poles as shown. Today flexible poles are allowed, as shown in the photograph on the opposite page. (1964)

Dave Wilson, when a student at MIT in 1973, held the intercollegiate New England pole-vaulting record. Here he is seen practicing at the MIT Athletic Cage. He set up the strobe lights and camera and directed Professor Edgerton and a student on how to aim the lights and when to expose the pictures.

Rodeo Rider

Bernice Dossey rides standing on the saddle. (1941)

Home Brew

Cecil Henley is flying from his horse, Home Brew, during a Boston Garden performance of the rodeo. The flash tube was located in a 30-inch searchlight reflector in the ceiling, and the axis of the reflector was continuously pointed at the action. (1940)

Song and Dance

Seven Photographs by Gjon Mili

Edith Piaf, singing "l'accordeo-
niste" in Paris (1946)

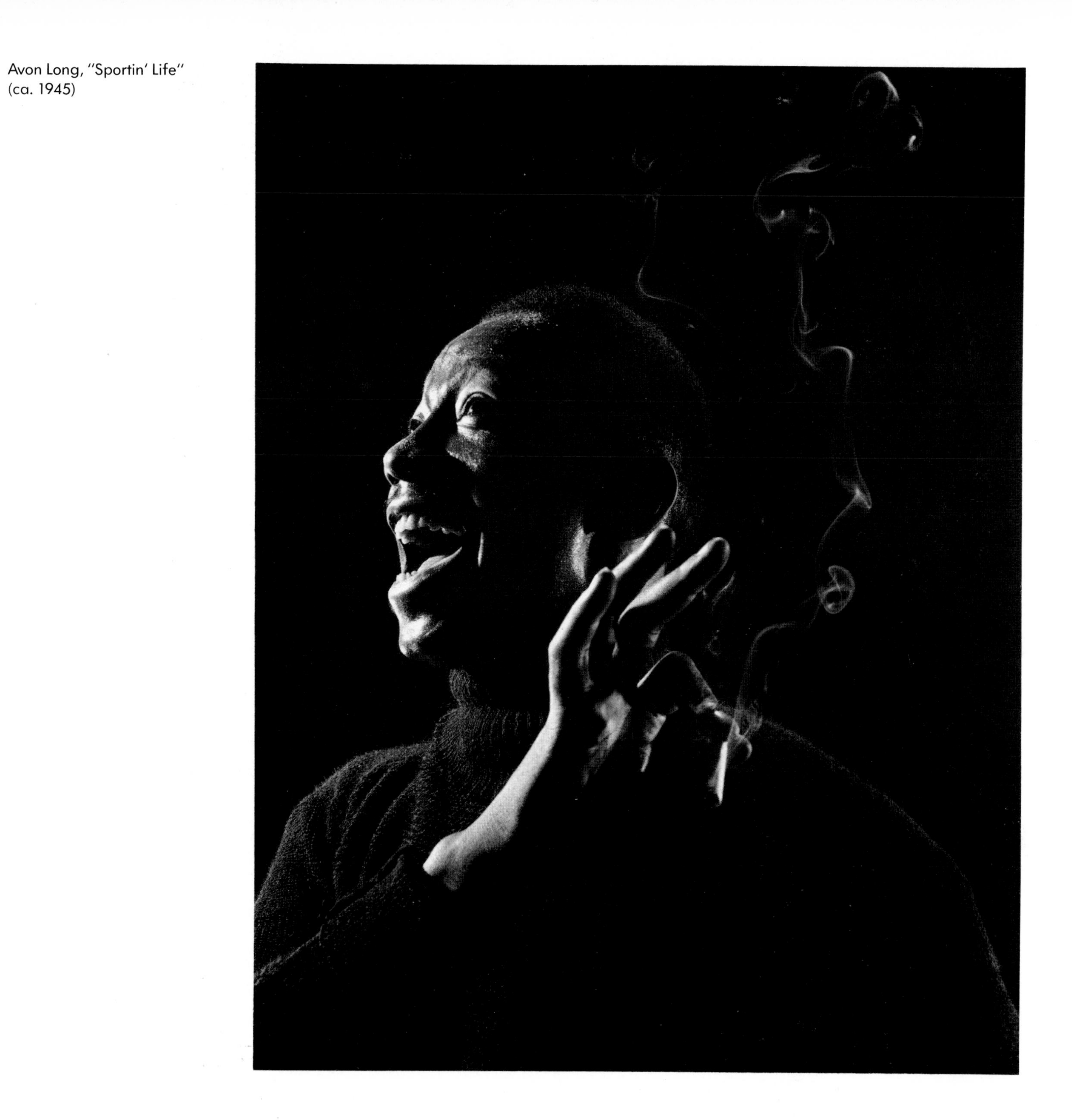

Avon Long, "Sportin' Life"
(ca. 1945)

Moune de Rivelle being drawn
by Miguel Covarrubias (ca. 1947)

Gene Krupa, "Drum Beat"
(ca. 1941)

Betty Bruce, tap dancer
(ca. 1941)

Tanaquil Le Clercq (ca. 1952)

"Centaurs" — Charles Wideman, Jose Limon, Lee Sherman (ca. 1939)

The Poetry and Joy of People in Action

Gus Solomons Dancing

Multiple-flash photographs of dancer Gustave "Gus" M. Solomons, Jr., in striking Shiva-like routines. These photographs were made when Solomons was a student member of a dancing group at MIT.

David Epstein Conducting MIT Orchestra

Charles E. Miller

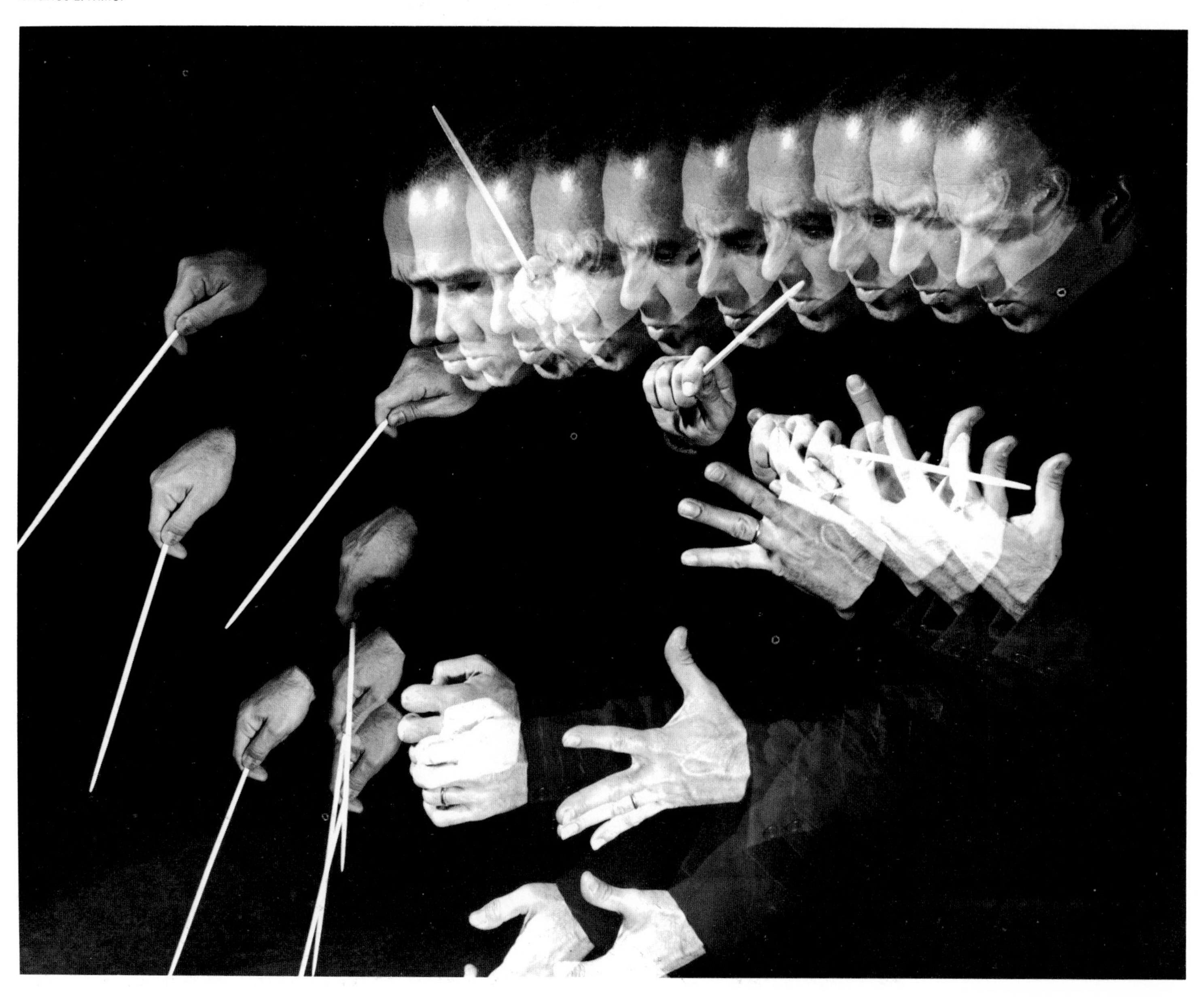

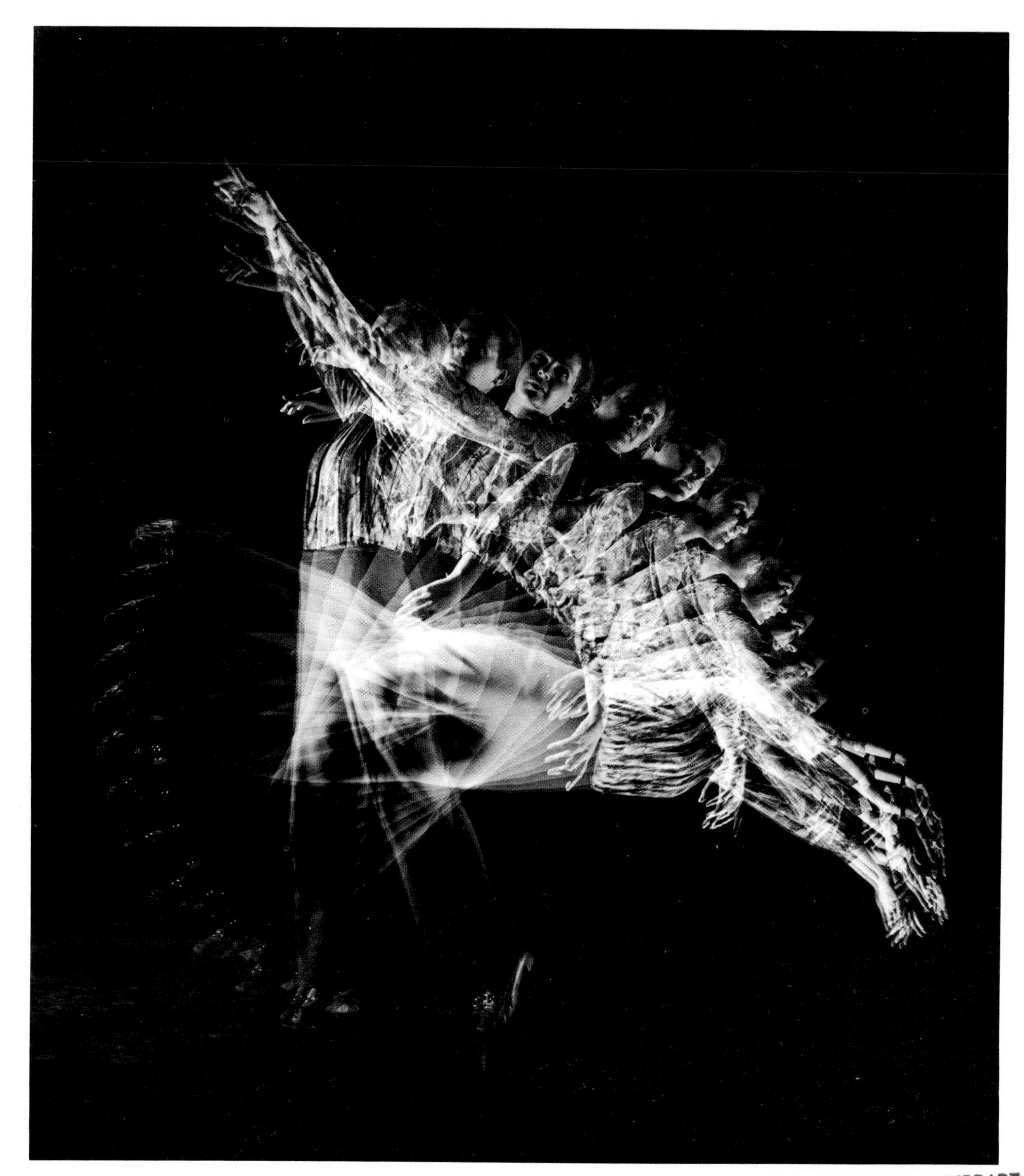

Seeing a Dancer by Multiple Flash with a Moving Camera (1971)

George Clemmer

Drum Majorette

In stroboscopic light flashing at 60 times per second, Muriel Sutherland, a drum majorette at Belmont High School, Massachusetts, twirls a baton, left, and grasps it as it falls, right.

Skipping Rope

Flowing Motions

Multiple-exposure photographs of people in flowing motion. The above picture was taken at the rate of 10 exposures a second, the opposite pictures at the rate of 30 exposures a second.

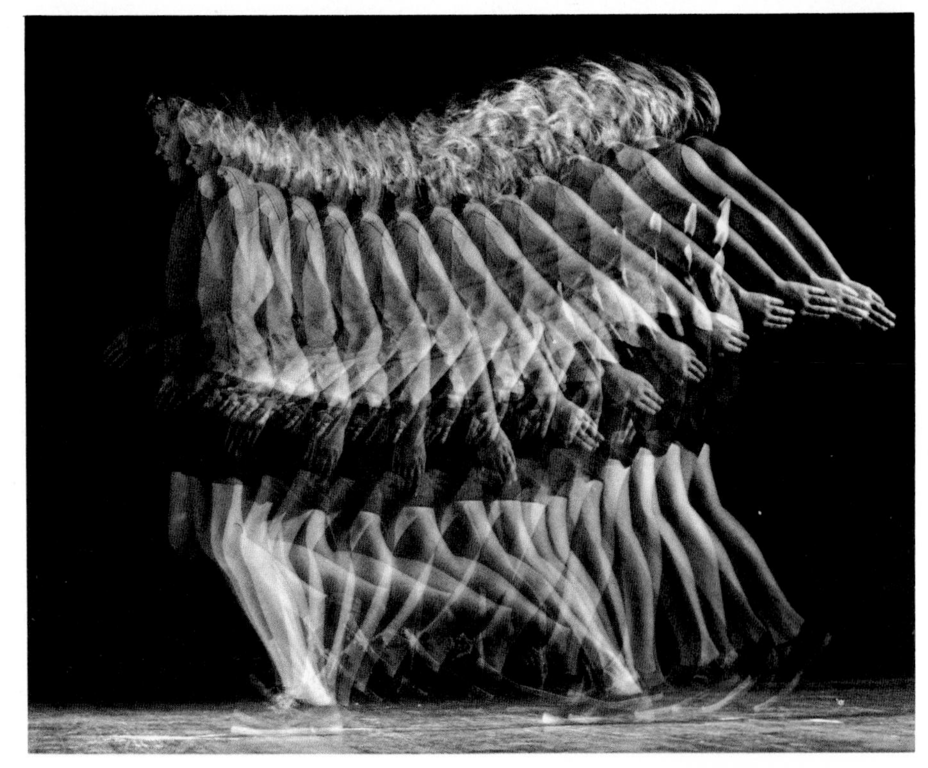

A Back Dive

Only Five Balls

Jerome "Skip" King, member of the MIT Juggling Club, keeps five softballs in action. The photograph consists of ten strobe exposures taken 1/30 of a second apart.

Papa Zacchini Shoots His Daughter

In its heyday this was a famous circus stunt. Edgerton caught the flight of the lady, Victoria Zacchini, as she was projected from the cannon. She ultimately landed safely in a net after her 175-foot trajectory.

The Zacchinis' equipment for this famous act is now enshrined in the Circus Museum at Sarasota, Florida.

What Would William Tell Think?

An apple shot with a bullet traveling at 900 meters per second. Photographed with a microflash stroboscope at an exposure of one-third of a microsecond. (1964)

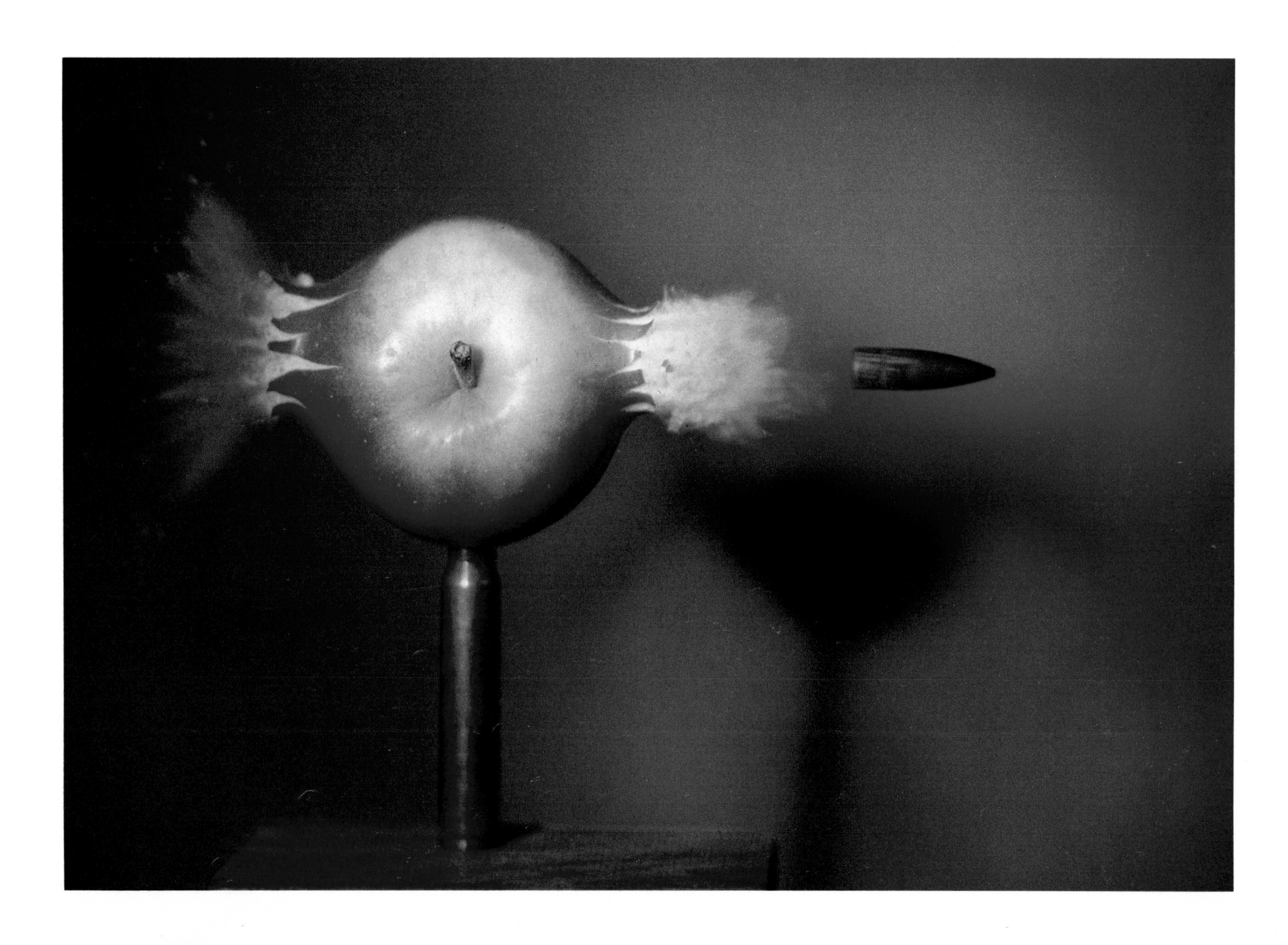

Supersonic Candlelight

A stroboscopic color schlieren or shadow picture taken at one-third microsecond exposure shows a supersonic bullet passing through the hot air rising above a candle. Schlieren pictures make visible the regions of nonuniform density in air. The stroboscope has given new capabilities to shadow photography. (1973)

Kim Vandiver and Harold Edgerton

Fan Blade Vortex

At the tip of a fan blade is a vortex formed by heated air above an alcohol lamp. This stroboscopic color schlieren taken at one-third-microsecond exposure was made with a multi-color source of light. (1973)
Kim Vandiver and Harold Edgerton

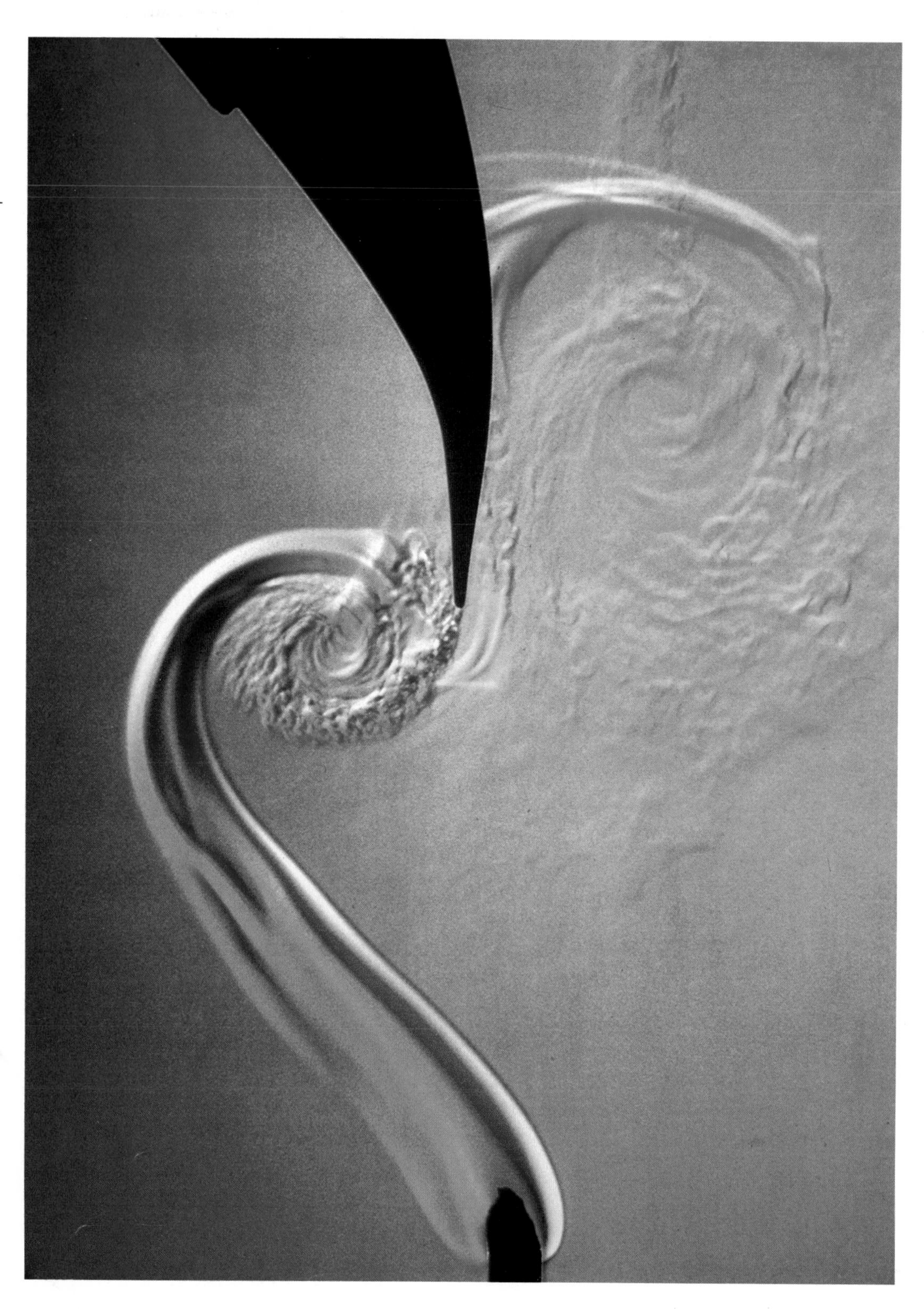

Peruvian Hummingbird
(Loddigesia mirabilis)
Native to the Andes of northern Peru, the bird is flying backward from the perch as indicated by the bend in the central pair of rectrices; the bird has controlled the outer pair for this particular maneuver.

Photograph by Crawford H. Greenewalt

Brazilian Hummingbird
(Topaza pella)

Topaza pella is one of the most spectacular hummingbirds but very difficult to see because it lives in dense jungles in the Amapá province of Brazil.

Photograph by Crawford H. Greenewalt

Grass Green Tanager
(Chloronis riefferi)

This male in flight was found near Baños, Ecuador. Bleitz, an engineer, author, ornithologist, and naturalist, is president of the Bleitz Wildlife Foundation, Hollywood, California. He has color photographs and descriptive material covering over 650 species of birds.
Don Bleitz

Ballet of the Guppies

Two fancy male guppies *(Poecilia reticulata)*, photographed swimming between two underwater mirrors adjoined at 90°, to produce multiple images. Single strobe light from above.

Paul A. Zahl, scientist, editor, author, and Senior Natural Scientist of the National Geographic Society, has held scientific appointments of distinction in a number of universities and research institutions.

Paul A. Zahl,

Tube Worms on Ocean Floor

In the near-freezing depths of the Galapagos Rift warm springs or hydrothermal vents have been discovered on the sea floor. Around these warm, sunless springs a National Geographic expedition found and photographed with stroboscopic lights marine life never before seen or known.

John M. Edmond,
©National Geographic Society

Reddish Octoral and Brittle Star Near Sunless Springs

Jack R. Dymond,
©National Geographic Society

Moon Jelly *(Aurelia aurita)*

Kenneth R. H. Read

Mysterious Poetry of Underwater Photography

This rare photograph was made in the Red Sea with an underwater xenon flash lamp and daylight illumination. The fish would not have been exposed without the flash.[7]

David Doubilet,

Gymnast Cathy Rigby on the Balance Beam

Phillip Leonian, in recent years, has specialized in the photographic description of movement. "I have returned to the walking figure constantly," he has written, "whenever I have questions about ways to see." In 1978 the Neikrug Galleries in New York City presented an exhibition of his photographs entitled *Phillip Leonian's Walking Show*.

Rushing Girl

This strobe photograph by Ben Rose expresses Edgerton's "let's get going" drive and energy.

Ben Rose has been president of the American Society of Magazine Photographers and at present teaches a course at Parson's School of Design. His primary commitment, however, is the development of methods and equipment for extending the camera's ability to portray movement. Rose is noted as an advertising photographer who has constructed special electronic devices to achieve unique photographic effects. He is able, for example, to make multiple stroboscopic shots in bright sunlight without interference from existing or ambient daylight.

Ben Rose

Photographing the Sea's Dark Underworld

In an article published in 1955 in the *National Geographic* magazine, Edgerton recalled the following passage in Jules Verne's *Twenty Thousand Leagues Under the Sea,* published in 1869:

"Look, Captain, look at these magnificent rocks, these uninhabited grottoes, these lowest receptacles of the globe, where life is no longer possible! What unknown sights are here! Why should we be unable to preserve a remembrance to them?"
"Would you like to carry away more than the remembrance?" said Captain Nemo.
"What do you mean by those words?"
"I mean to say that nothing is easier than to take a photographic view of this submarine region."

As usual with Jules Verne's science fiction, realization of his prophecies prove hard to turn into reality, but thanks to developments by Edgerton and others, it is now possible to explore the "dark unfathomed" depths of the sea—and to show they are inhabited.

Edgerton first became attracted to possible underwater uses of his equipment when he was approached by the late E. Newton Harvey, professor of biology at Johns Hopkins, who wanted pictures of phosphorescent fish that live thousands of feet deep in the ocean. This goal had not yet been accomplished because cameras were not available that would withstand deep sea pressures. But Edgerton was inspired to develop a strobe underwater camera, and with the assistance of staff at the Woods Hole Oceanographic Institution, he succeeded. Now Edgerton-type cameras, along with others, are available to take pictures in waters seven miles deep. New companies such as Benthos, Inc., now manufacture Edgerton deep sea photographic systems for use at any depth.

A second event led Edgerton not only into underwater photography but into underwater exploration and oceanographic research. Jacques Cousteau had approached the National Geographic for funds to support his newly acquired *Calypso*. He also wanted to study the deep scattering layer in the ocean that moves up to the surface at night and moves down when morning comes. Cousteau wanted to find out why and suspected that biological activities in or about the layer might explain this strange phenomenon. Edgerton devised equipment to study the rising and falling layers. The pictures revealed no significant evidence that biological activity affected the up and down movement, but this experiment led to a partnership between Edgerton and Jacques Cousteau in which Edgerton made ten trips on Cousteau's ship, the *Calypso*, and participated in searching under water for ancient sunken vessels and other archaeological artifacts in the Aegian Sea. He also contributed his experience and equipment to the exploration of sea floors in the Mediterranean, the Atlantic, and Lake Titicaca in Bolivia. When Cousteau embarked on his search for the sunken liner *Britannic*, sunk by a mine in 1916 off the Greek island of Cos, he called upon Edgerton to use side-scan sonar to locate the sunken ship. And later, Edgerton was also enlisted to help in the search for the wreck of the Civil War *Monitor*.

Through these and other oceanographic searches, Edgerton became deeply involved in oceanography, and other explorers, naturalists, and archaeologists sought him out to assist in their underwater work. The late Carl J. Shipek, oceanographer with the Naval Undersea Water Center at San Diego, wrote to Edgerton that "The United States is a leader in ocean exploration because of your achievements." Four members of the Institute of Oceanology, of the Soviet Academy of Sciences, including its director and deputy director, wrote, "Your contribution to the progress of oceanography is well known and has won deserved recognition among all those who are sincerely concerned about its further advancement to the benefit of people."

In a private letter, Melvin M. Payne, president of the National Geographic Society, celebrated Edgerton's contributions to marine photography. "Looking back over the years, I recall, too, the day that Luis Marden returned excitedly from a meeting with you in Cambridge. During that 1951 visit you had fired his imagination—and shortly thereafter our Research Committee's interest—in the construction of an undersea camera and high-speed flash unit that could be lowered to great depths. Often as I have wandered past your early camera and flash equipment in Explorers Hall, I have thought how far we have progressed in marine photography in so short a time as a result of your vision and engineering brilliance."

The pictures that follow, as well as the underwater pictures in the colored section, are examples of revealing pictures that show the nature of the sea floor and the activity in the deepest parts of the ocean.

Portrait of an *Abyssal enteropneust*

The late earth scientist Maurice Ewing sent Edgerton this electronic flash photograph of a deep ocean worm, or acorn worm, as it discharges its excrement of mud on the ocean floor. While such tracks are common on the ocean floor, this picture, as far as can be determined, is unique because it shows the worm.

Lamont Laboratory, Columbia University

Flying Fish

This photograph, composed of two superimposed photographs taken near Avalon Bay, Catalina Island, California, shows the upper fish in full flight. The lower fish is about to "take off" as he gains speed by violently zigzagging his tail in the water.[8]

Ocean Floor

A section of the ocean floor as viewed by an underwater camera

Richard Pratt
Woods Hole Oceanographic Institution

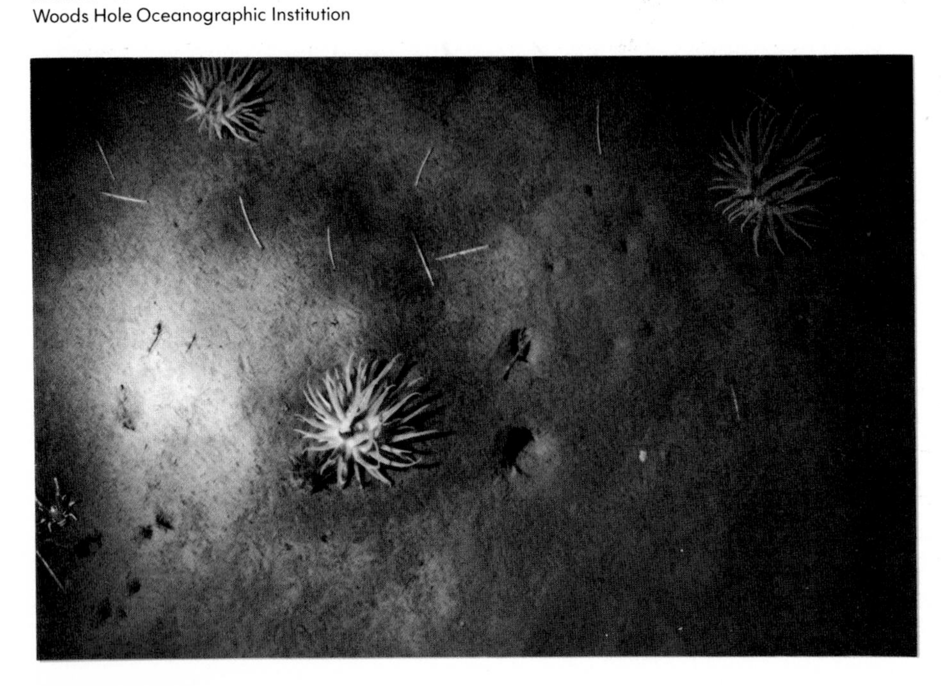

Sea Cucumbers and Brittle Stars

The strobe light illuminates giant sea cucumbers and brittle stars at a depth of 4,000 feet from a Westinghouse submersible Deepstar.

Donald Keach

Siphonophore

Rare photograph of a siphonophore taken in the Mediterranean. Edgerton spent the good part of the summer of 1954 aboard Cousteau's *Calypso* taking rare pictures of these and other hard-to-photograph marine organisms for an understanding of the deep scattering layer.

School of Codfish

In a study of the seafloor environment of a trans-Atlantic cable, this school of codfish was observed swimming at a depth of 1,155 feet.

G. R. Leopold,
Bell Telephone Laboratories

In an Underwater Garden

In an early attempt to photograph the ocean floor, Edgerton and the staff of Cousteau's *Calypso* achieved this view by electronic flash. (1954)

Finding The Sunken *USS Monitor*

During a stormy December night in 1862, the famous Civil War ship, the *USS Monitor*, sank while under tow off Cape Hatteras, North Carolina. Over the years many searched for the *Monitor* without success, but in 1973 a group headed by John Newton of the Duke University Marine Laboratory located the sunken ship. Edgerton participated in this successful effort by bringing to bear his experience and equipment for side-scan sonar search and underwater photography.

Once the wreck was located, an Edgerton underwater camera became snagged in the wreck and could not be retrieved. The photograph shows the camera caught in the wreckage at the right. In later expeditions Edgerton made a determined effort to retrieve it, hoping that it might still have useful film that would give further information about the *Monitor*. An expedition in 1977 was successful in retrieving the camera assembly, but the film in it had been ruined by the immersion in the sea. The recovered camera is now exhibited in Edgerton's museum, Strobe Alley at MIT.

The encrusted side armor of the *Monitor* and schools of fish

Montage of section of the sunken *USS Monitor* made up of photographs taken with stroboscopic light from the *Alcoa Seaprobe*. The *Monitor* was difficult to identify at first because it was upside down on the sea bottom, and the turret was broken off the hull and below the overturned ship. Positive identification was possible from this and other photographs and from videotapes.
Glen Tillman

Hippocampus

In the first picture, the sea horse studies the air bubbler that supplies oxygen to his water. Although the sea horse is slow in his motions, his fins vibrate quickly, about 35 beats per second.[9]

Medusa

One of several types of underwater cameras designed in the Stroboscopic Laboratory at MIT is the luminescence camera of Lloyd Breslau for investigating marine bioluminescence. These organisms trigger the camera with luminescent flashes and thus take their own pictures.

Here is a medusa in the Gulf of Mexico that triggered its own picture.

Lloyd Breslau

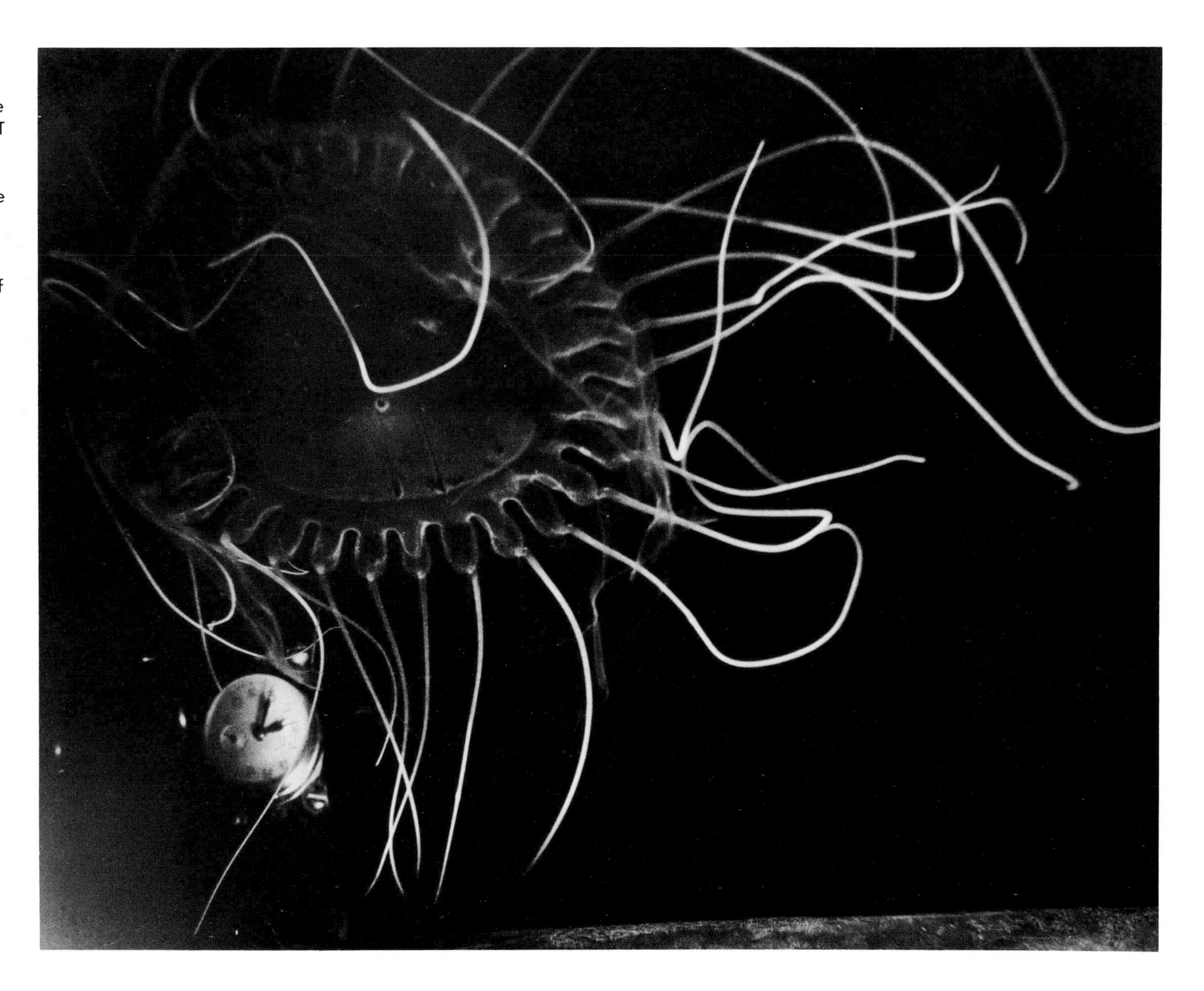

For Deep Sea Photography

Two Edgerton-type cameras are at the top of the structure and will produce a stereo pair of photographs. Each has 800 35-mm. exposures and a special lens corrected to operate through water.

A 200-watt-second strobe light is located near the bottom of the unit to minimize the distance that the light travels in the water.

Midway in the gear is a "pinger" that enables the operator on the ship to keep track of the camera-to-bottom distance and make suitable corrections with the camera-lowering mechanism.

This equipment was used by the Woods Hole Oceanographic Institution to survey the wreckage of the *Thresher* submarine lost in deep water east of Boston during a trial run.

Atomic Bomb Explosions

On the following pages are sample nonstrobe photographs of atomic bomb explosions which are intensely luminescent. While not so dramatic or photographically clear as the others, the first two are of special interest. They show the nuclear explosion before the little house or cab enclosing the bomb had been vaporized. Edgerton took the photographs from the top of a seventy-five-foot tower seven miles away. A magneto-optic shutter was used for the split-second timing, and extraordinarily delicate arrangements were made to focus and direct the camera from so great a distance.

A Newtonian-type telescope with a ten-foot focal length was used to image the bomb. A relay lens then magnified the picture to about twice its size.

Alignment of the telescope and focusing were complicated by the mechanical deflection and vibration of the tower itself. A satisfactory method of focusing was evolved by the use of a 1,000-watt lamp placed at the position of the bomb. A vibration pattern appeared on the ground-glass image of the camera, but the focus was adjusted until the width of the lines of the pattern were minimized. Fission products released by the explosion made the scene fluorescent.

These and other photographs were taken automatically with a magneto-optic shutter by EG&G.

130–131

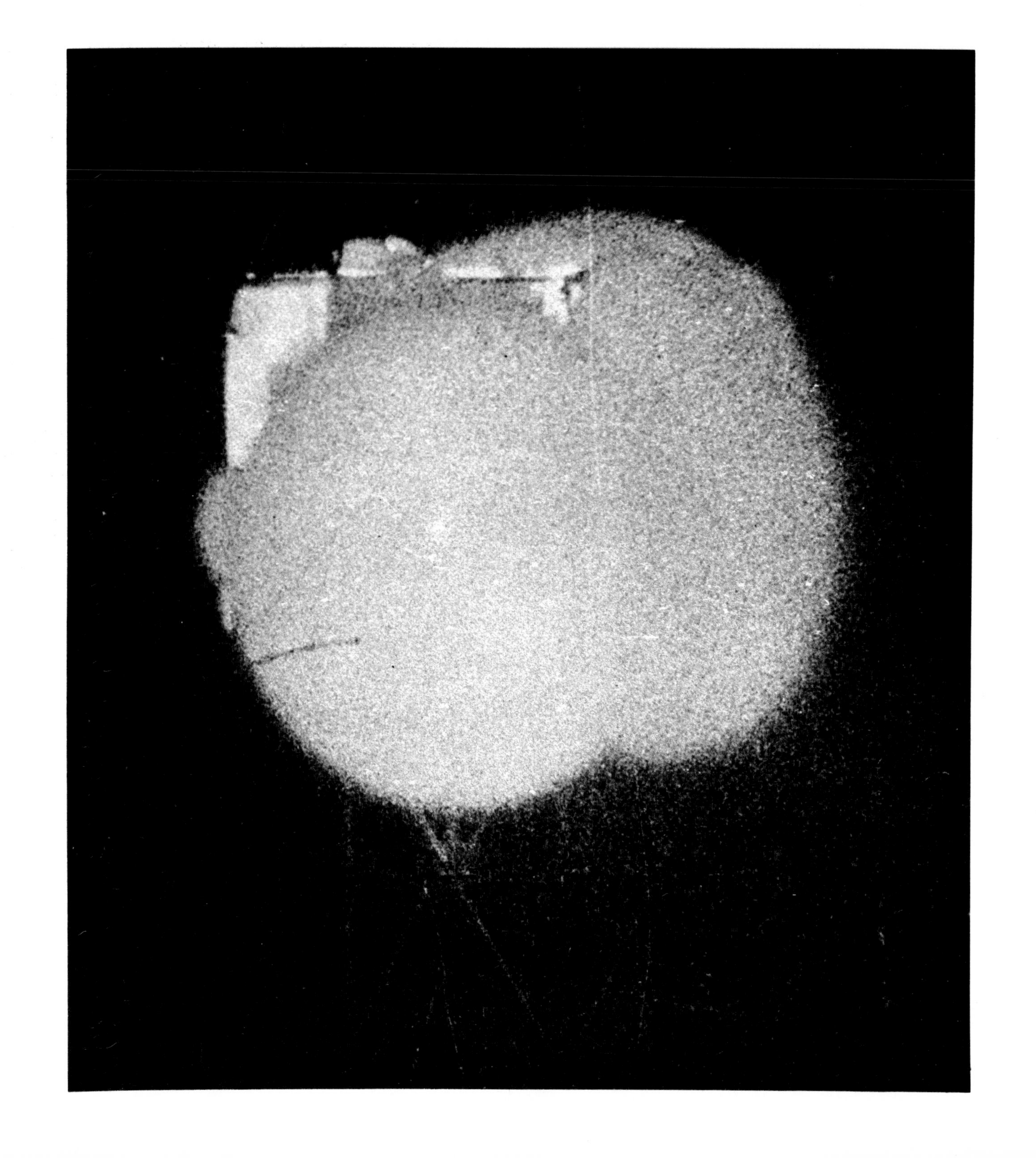

134–135

Photographic Reconnaissance at Night

It was inevitable that Edgerton's brilliant lights should be called into service for air reconnaissance in World War II and that Edgerton himself should be recruited to build the flash equipment and to train the crews for night photography.

At the behest of Colonel George Goddard of the Army Air Force, Edgerton went to Italy and put his night photography to use as the Allied Armies sought to break out of the Cassino stalemate. Fortunately, his MIT colleague, Edward Bowles, was an adviser to Secretary Stimson and arranged for Edgerton to be appointed as an officer of undesignated rank. This was important because it enabled Edgerton, as he has laughingly reported, to assume on occasion a rank one degree higher than any regular officer who might impede his crash program. Two planes with crews trained by Edgerton flew eighty missions in Italy to find out how the Germans were renewing their ammunition supplies while the Allied Armies sought to speed the German retreat up the Italian peninsula. After three months of successful operations in Italy, Edgerton was asked to go to England to build a squadron to use his equipment in land operations over Europe and in submarine search missions. The pilots who were assigned to this squadron were "hot" pilots who preferred fighter missions and who felt demeaned by being assigned to photographic missions. With an adroitness that became legendary, Edgerton set about to develop the interest of the flyers assigned to the reconnaissance squadron. In ways he has not explained, he obtained the coordinates of a nudist camp in England and included it among the targets to be photographed, giving a fillip to the otherwise boring training flights. He also used Stonehenge as a model for demonstrating the effectiveness of his flash lights. Edgerton recalls that he was on the ground when the plane flashed its light over Stonehenge and the camera he had fixed on a fence post photographed the ancient megaliths.

The Edgerton reconnaissance system yielded valuable information in regard to German troop movements on the Continent. When Patton and Bradley were making some of their great advances through France, intelligence inputs and the reconnaissance data from Edgerton's night photography indicated that there were no German forces in their way at critical places.

The success of flash photography in the European theatre led to the use of Edgerton's system in the Pacific. In May 1945, an Army Air Force officer reported that

> In Northern Burma . . . an F-7 with an Edgerton light unit was used to photograph Japanese movements. During its operations a Japanese officer was captured who could speak excellent English. He quoted his commmanding officer as saying the following about the F-7's operations:
>
> "Oh, what can we do now! with his bright blinking eyes streaking across the dark canopy of night, the devil himself has compromised our last and now unfaithful mistress of security."
>
> The importance of Night Reconnaissance cannot be overemphasized and its effect upon the enemy's morale cannot be overstressed, let alone the military importance of the information obtained through these Night Photographs.

For his great contributions to night reconnaissance, Edgerton received the Medal of Freedom, with the citation:

> Mr. Harold E. Edgerton, Civilian Technical Advisor, rendered meritorious achievement in connection with military operation in the development and installation of equipment for night photo reconnaissance during the period 1 June 1944 to 20 November 1944. His untiring effort, resourcefulness and competence made aerial night reconnaissance of enemy held territory a reality under adverse weather conditions. Without his invaluable knowledge, unswerving zeal and ingenuity under adverse conditions, this type of military operation would not have been possible at such an early date. The results of his endeavor have provided the U. S. Army Air Forces and Ground Forces with vital intelligence information that previous equipment of this nature was not capable of obtaining.

D-Day

The Edgerton airborne photographic equipment was intensively used in planning for World War II D-Day. This aerial photograph of a road intersection in Normandy was taken during the early morning hours of June 6, 1944, prior to the invasion. It revealed an absence of German forces at this important location. The diagonal streak at the bottom center of the photograph was light from a tracer machine gun bullet.

U.S. Army Air Force

Capitol

The Capitol in Washington from 3,000 feet, as shown by an electronic flash photograph taken in 1946

U.S. Army Air Force

Pentagon at Night

Electronic flash night photograph of the headquarters of the U. S. Defense Department.

Taken by U.S. Army Air Force with flash equipment provided by Harold E. Edgerton.

Stonehenge

Two wartime photographs of Stonehenge at night taken by a single electronic flash light carried by a plane. The first was taken by Professor Edgerton standing on the ground. The second photograph was taken by the plane equipped with the Edgerton system.

While not apparent in this picture, except to Stonehenge experts, trilithon stones 57 and 58 with their top rock are shown in a fallen position in this 1944 photograph. These were raised after the war and are now in an upright position.

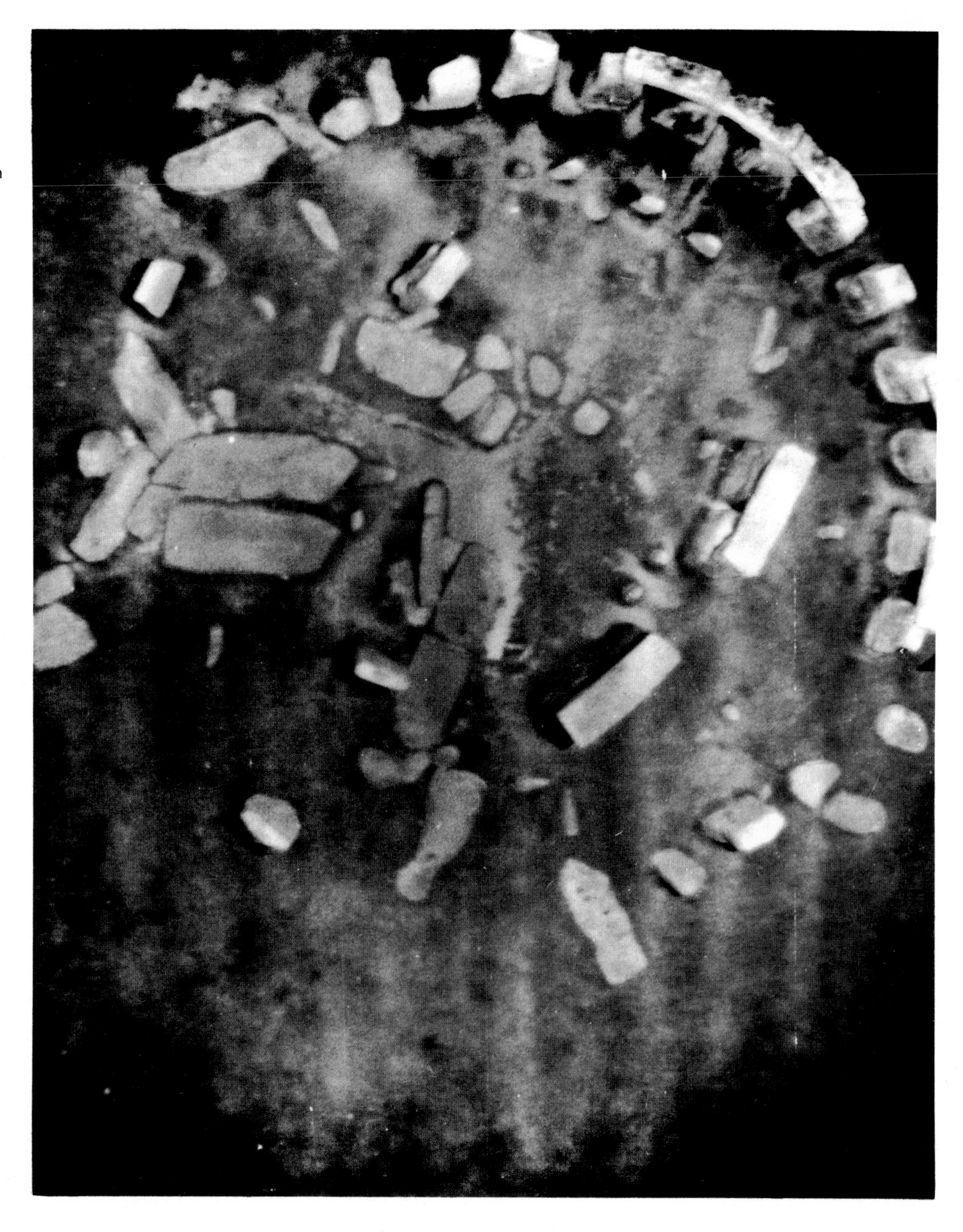

140–141

The Great Dome

MIT at night across the frozen Charles River Basin. Taken by students with the aid of Professor Edgerton, using 45,000-watt-second flash tubes and reflector. The camera was on the top of the Prudential Building, and the strobe light was on the roof of an apartment building on Beacon Street. A photoelectrically controlled strobe in Killian Court at MIT provided auxiliary lighting.

The Delectation of Art

Propagation of Glass Cracks

In these remarkable and beautiful moments of vision, a spring-driven metal plunger strikes the tempered glass plate with enough force to break it and starts an electrical timing circuit. At the proper split second the timing circuit sets off an Edgerton electric flash and exposes the negative for less than one millionth of a second. The timing is so accurate and responsive to control that a crack moving at nearly a mile a second can be stopped dead in its tracks at any desired point—as the two pictures show.

This high-speed photographic method of studying the propagation of glass cracks was carried out by Frederick E. Barstow, then a graduate student at MIT working with Professor Edgerton. Not only do the photographs record patterns of a beautiful, round dance, they also provide glass makers with useful information about the physical characteristics of their glass.

This is an example also of how a technical study using the strobe can also produce what André Malraux called the "delectation of art."

Among the optical paintings of the British artist, Bridget Riley, is a circular one she named *Blaze I* (1962). Its beautiful pattern suggests a somewhat similar pattern of the shards and cracks captured in these high-speed shots.

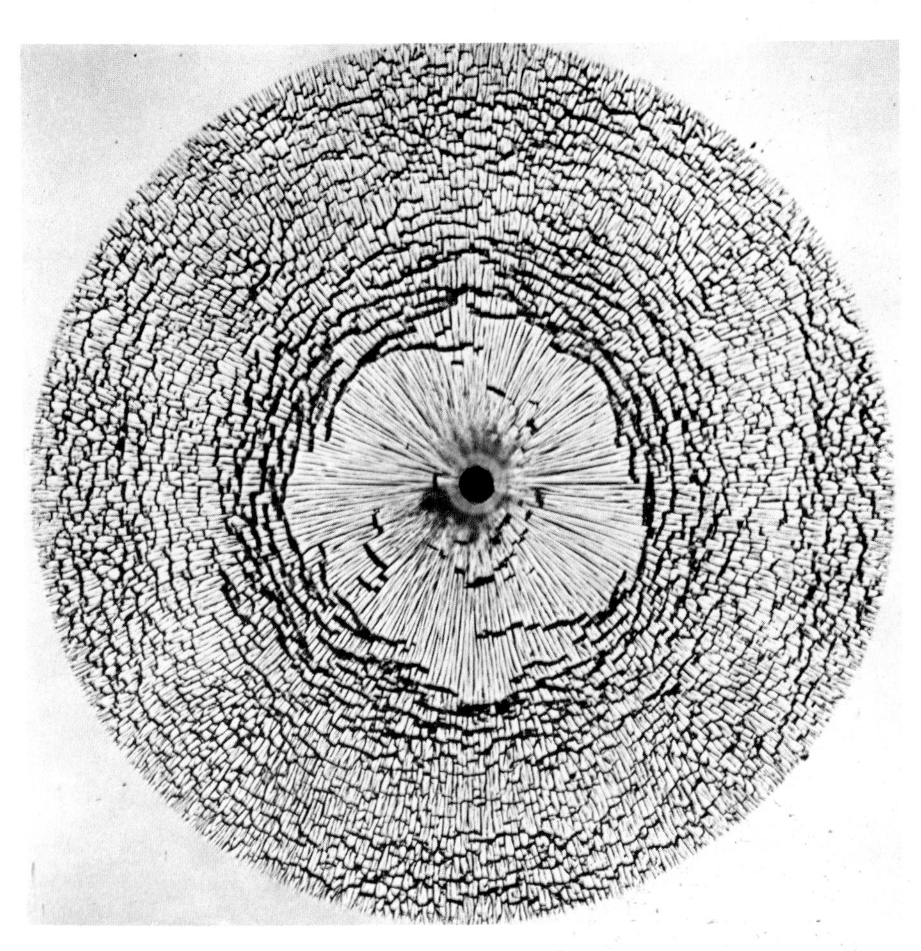

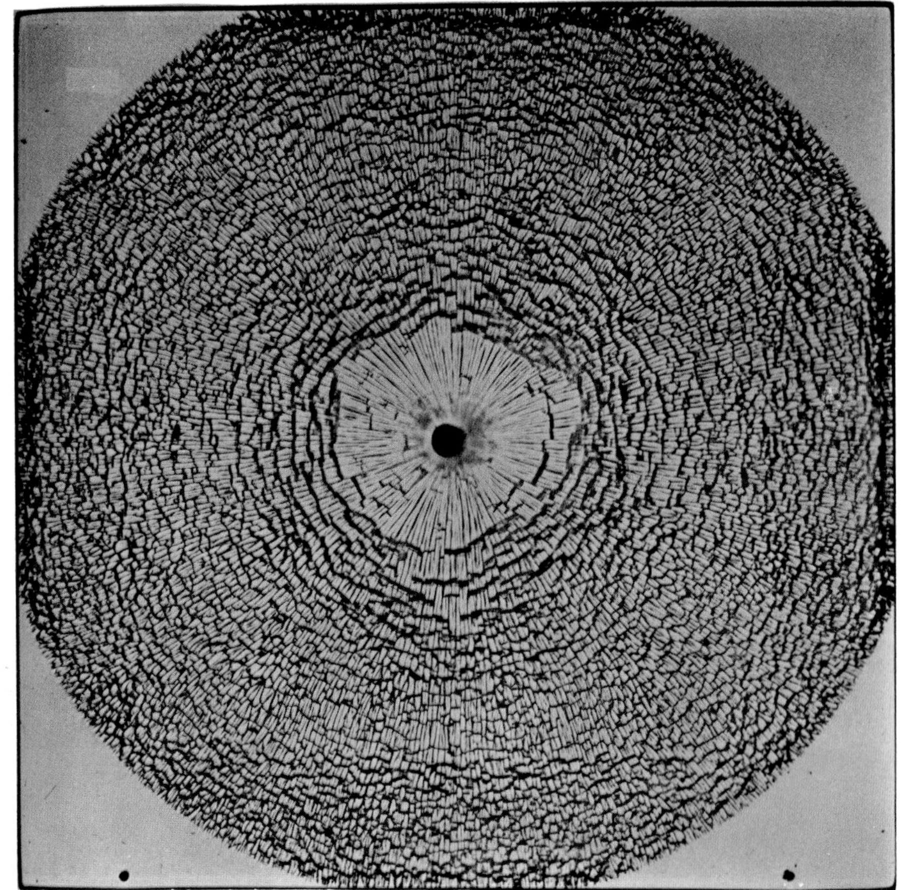

Hammer in Flight

Multi-flash photograph of a ball peen hammer in flight

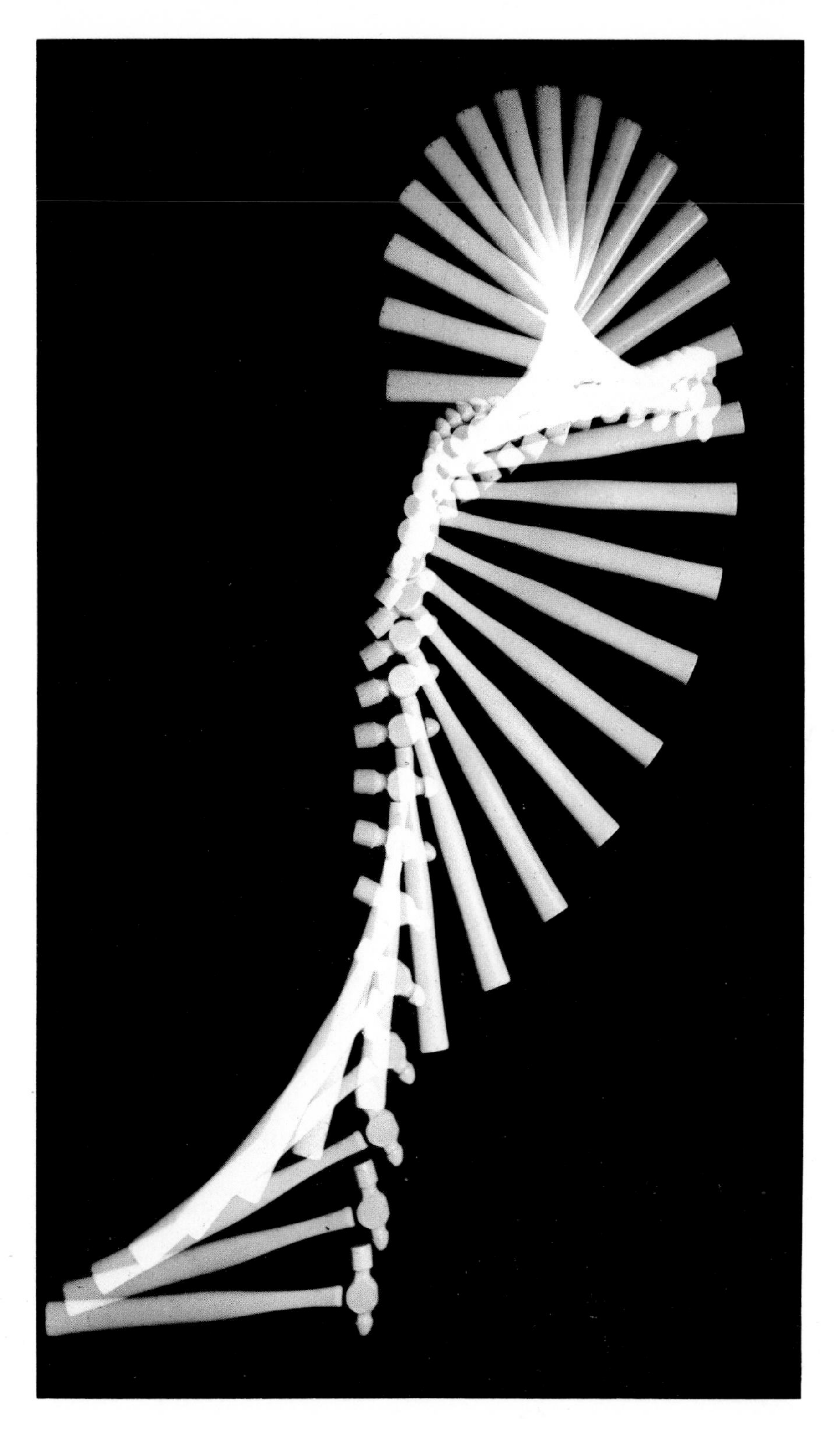

Shadow Photography Redivivus

As Used for Seeing and Studying Marine Organisms

Shadow or silhouette photographs were made as early as 1834 by W. H. Fox-Talbot, the remarkable pioneer of photography. He made shadow negatives of leaves and lace on photosensitive paper and then produced positive prints.

The Edgerton stroboscopic system with its brilliant flashing light revived shadow photography and made it possible to put it to effective use in recording tiny organisms present in ordinary water and the seas. The subjects are put directly upon fine-grain film in their normal water environment and then exposed with a very small strobe lamp at a great distance. Subsequent enlargement of the film is made to reveal details.

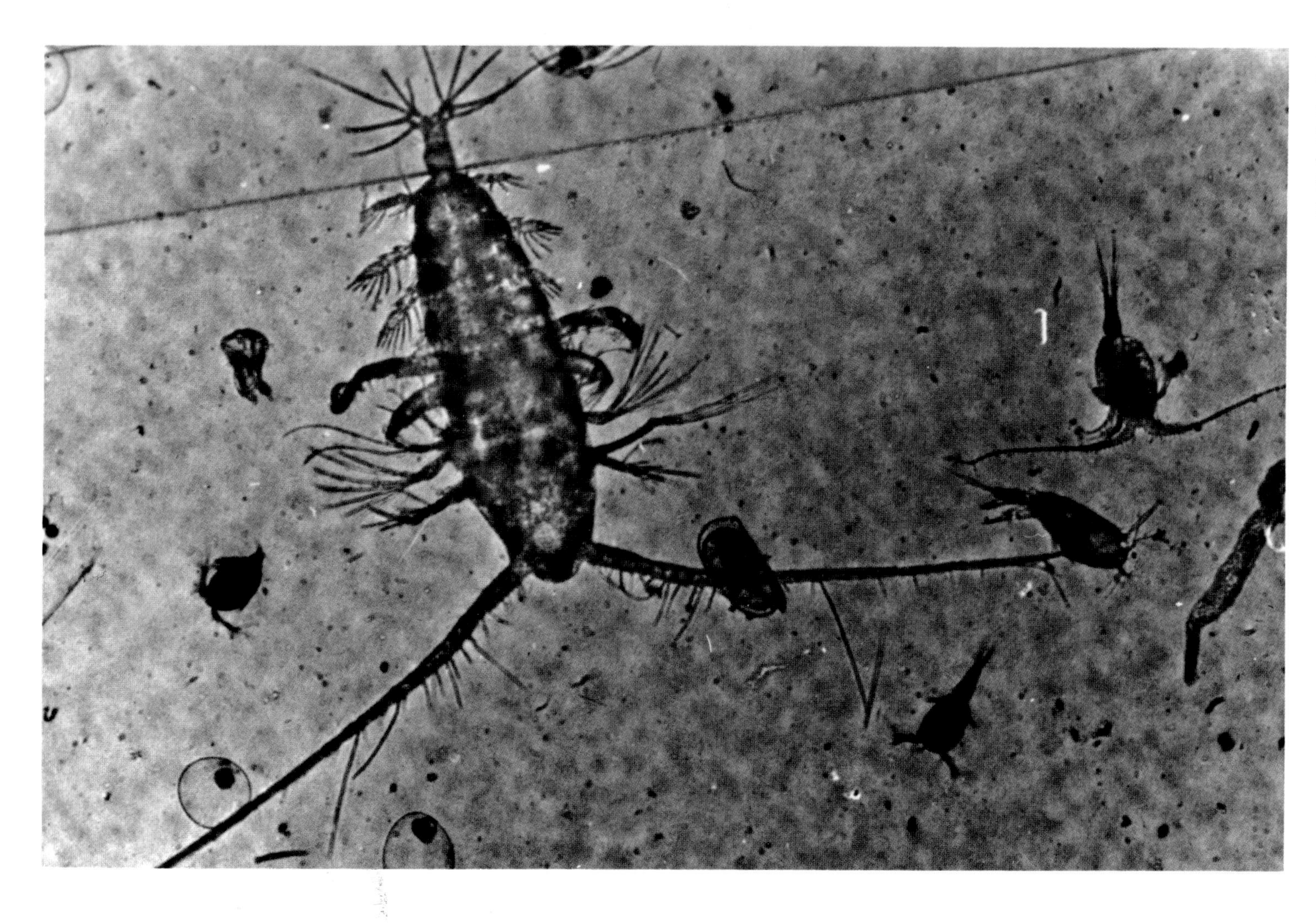

Variety of life in the amphitheater of the sea. In a moment of vision, silhouette photography provides a glimpse of copepods and eggs in the sea off Miami, Florida.
Peter Ortner

Plankton from Gulf Stream

Shadow photography reveals long, narrow marine worms (chaetognaths) and numerous copepods and ostracods (crustaceans) among plankton in this ocean water sample.
Peter Ortner

Live Plankton

Live plankton in Gulf Stream immediately after being netted from the sea. The large leaflike object is the larva of an eel, probably spawned in the Sargasso Sea.
Peter Ortner

Cirratulid

Photograph of a cirratulid, at the New England Aquarium, magnified about thirty times

Jeffrey Wilson,
in collaboration with Carolyn Karp

Newly Hatched Brine Shrimp and Eggs

A double exposure silhouette photograph made with two flashes of light about one-half second apart.

Note that the circular brine shrimp eggs are dark; since they do not move, they are exposed twice in the same place. The active brine shrimps are shown in two separate positions, each exposed by a single flash.

The velocity of the brine shrimp can be measured by noting the distance traveled between flashes and dividing it by the time interval between flashes.

Enlargement is about eight times.

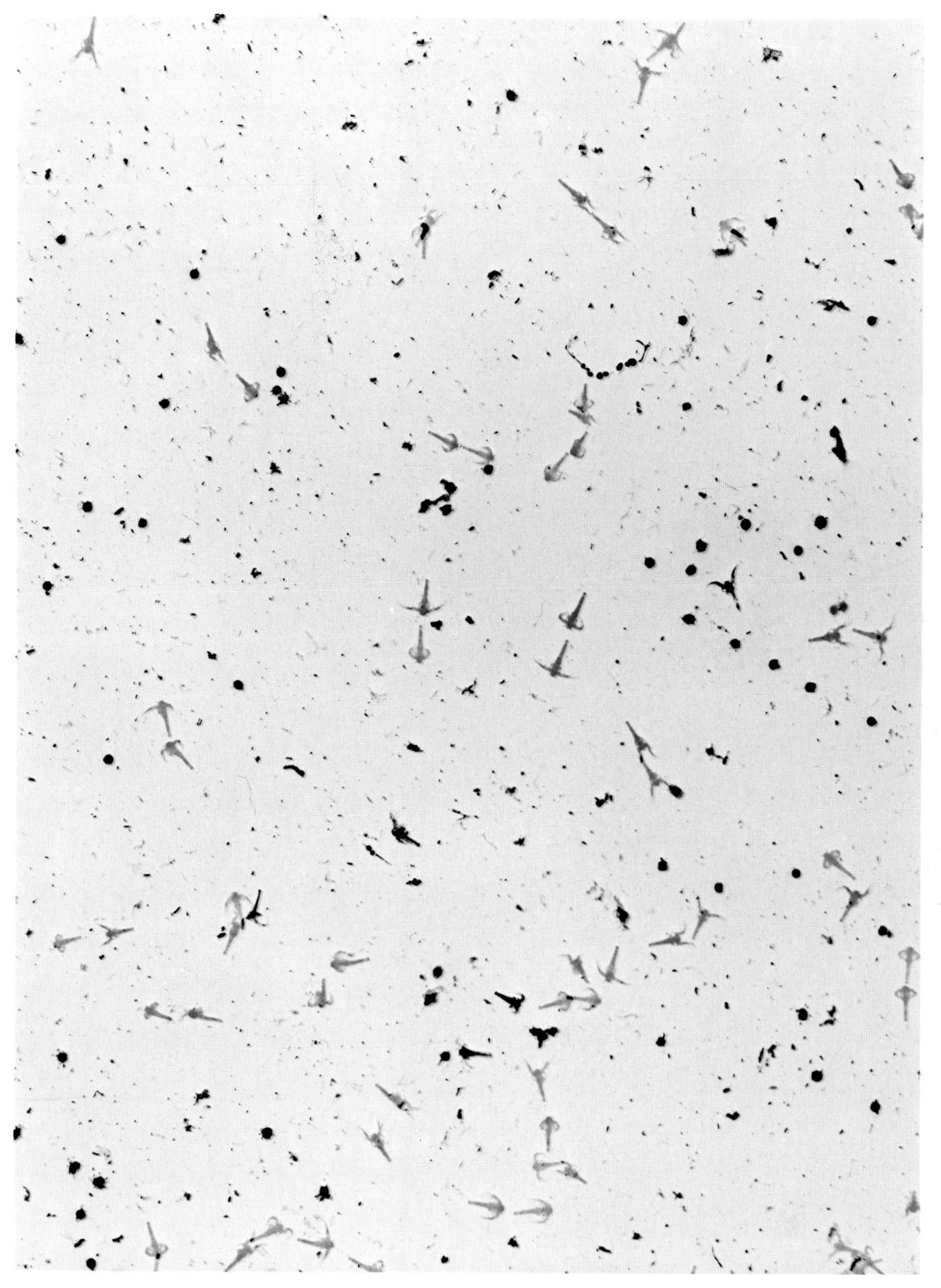

Brine Shrimp

Recently hatched brine shrimp photographed in action. Each shrimp is one millimeter long, having been hatched from an egg one-quarter millimeter in diameter. These sea animals are commonly used for fish food.

City Tap Water

Magnified ten times, the small black dots are particles in the water that we drink every day. What appears to be a string of beads is a hair floating on the water, probably fallen from the head of the photographer. The oily substance on the hair prevents it from being immersed. As it floats on the surface, it acts as a lens that deflects the strobe light.

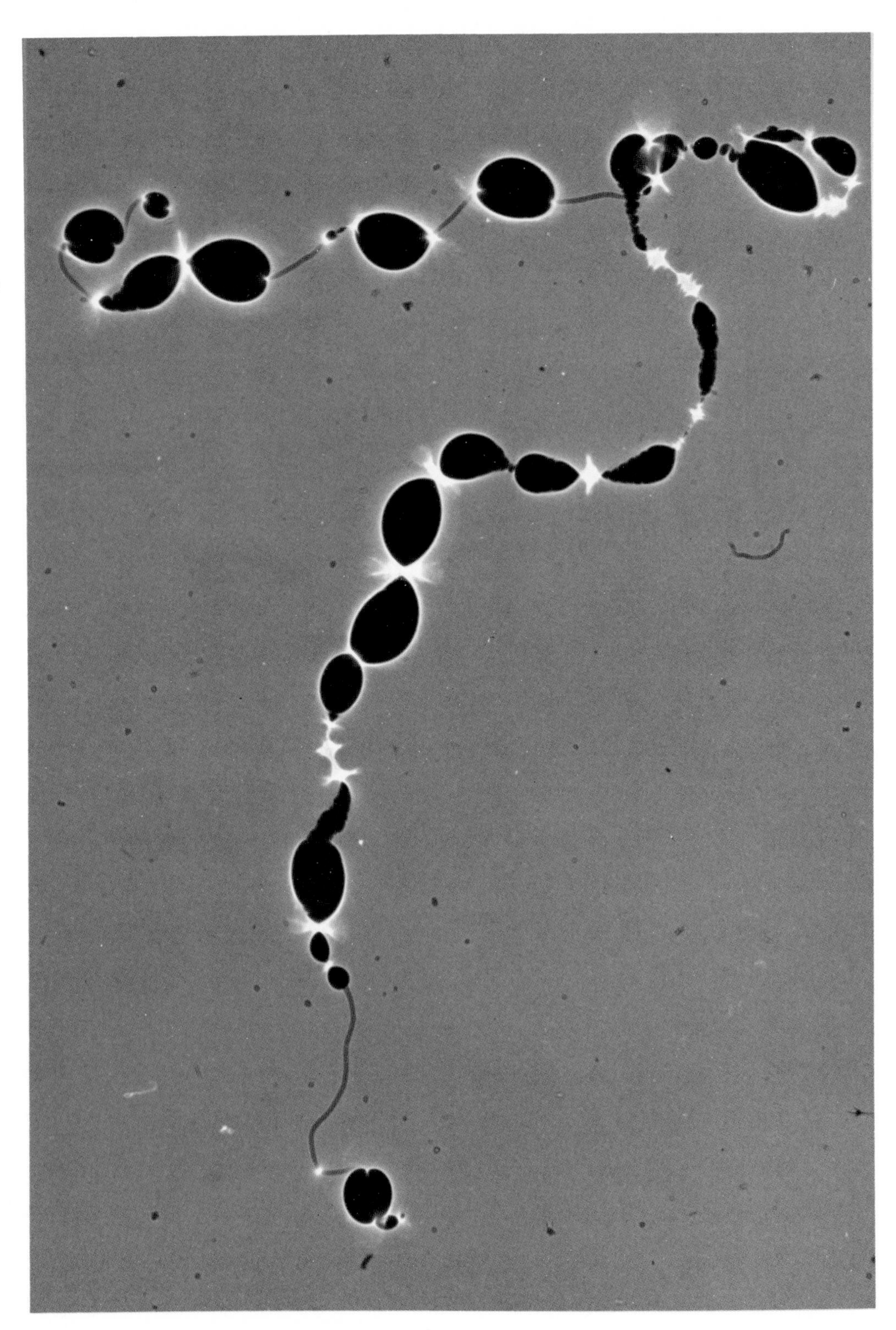

Sight with Sound

"Pingers," "Boomers," and "Fish"

As indicated in earlier portions of this book, Edgerton found it necessary to supplement his underwater cameras with sound systems because of the serious limitations caused by the properties of water or the absence of adequate light. Murky water frequently makes it impossible to obtain useful photographs. Even with the clearest ocean water, the maximum distance at which an object can be photographed is about 30 meters.

Penetration Sonar

The echo sounder is almost universally used today on ships to measure the depth of the water beneath them. With high peak power outputs for a short duration it is very useful for penetrating into the bottom of the sea. There are specialized applications for the echo sounder such as finding ancient ships long since submerged in mud or sediment. Similar but more powerful sound methods used for the discovery of oil and mineral resources have been given such names as "pinger," "mud penetrator," "boomer," "sparker," or "air gun."

An example of information yielded by penetration techniques is the sonar picture of the two vehicular tunnels that link Boston to its airport; these are about four meters below the bottom of the harbor. Another picture shows geological layers in the mud deposited in the late glacial age.

Side-Scan Sonar

Side-scan sonar is used in searching for underwater targets that project above the bottom of the sea. With such equipment a larger area can be surveyed more quickly than with a vertical beam echo sounder.

In his underwater search activities Edgerton uses a specialized type of side-looking, or side-scan, sonar to send a beam or beams of sound that can be projected from the survey ship or from a "fish" towed behind the ship.

Sound Views of the *USS Monitor*

Four side-scan sonar records of the *USS Monitor* wreck at Cape Hatteras, North Carolina, made aboard the *R/V Eastward*, August 1973.

The current (about 2 knots) has scoured the sand around the wreck to make a path that is visible to the sonar beam for 200 meters. It is believed that the flow of water around the wreck increases the velocity of the current that sweeps away the finer particles of sand. The surface then reflects sound more efficiently.

The turret of the *USS Monitor* lies underneath the inverted hull, at the stern. About half of the turret is visible. The sonar shows this turret marginally in the upper left-hand sonar diagram.

From these and other records, the position and attitude of the *Monitor* was found to be 29,954 meters to Cape Hatteras lighthouse, 19,265 meters to Diamond Shoals Tower, and almost due east/west with the bow to the west.

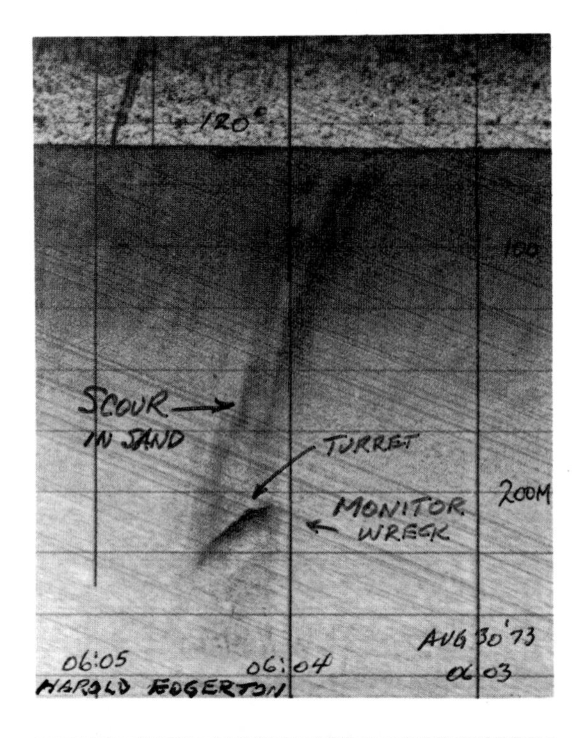

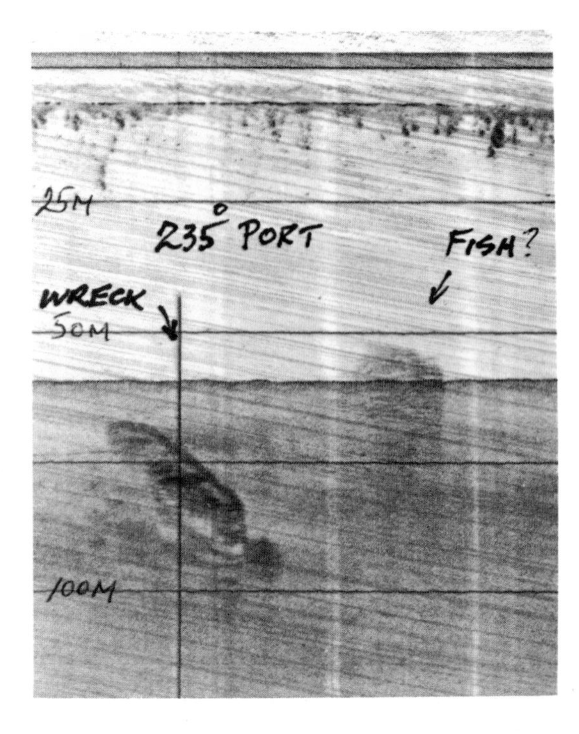

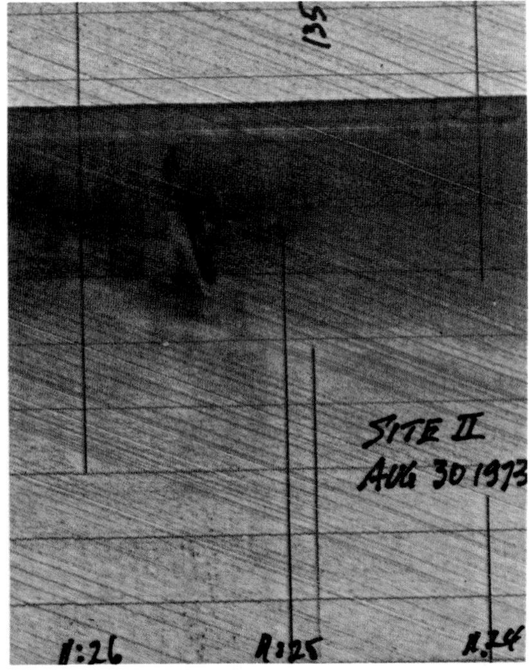

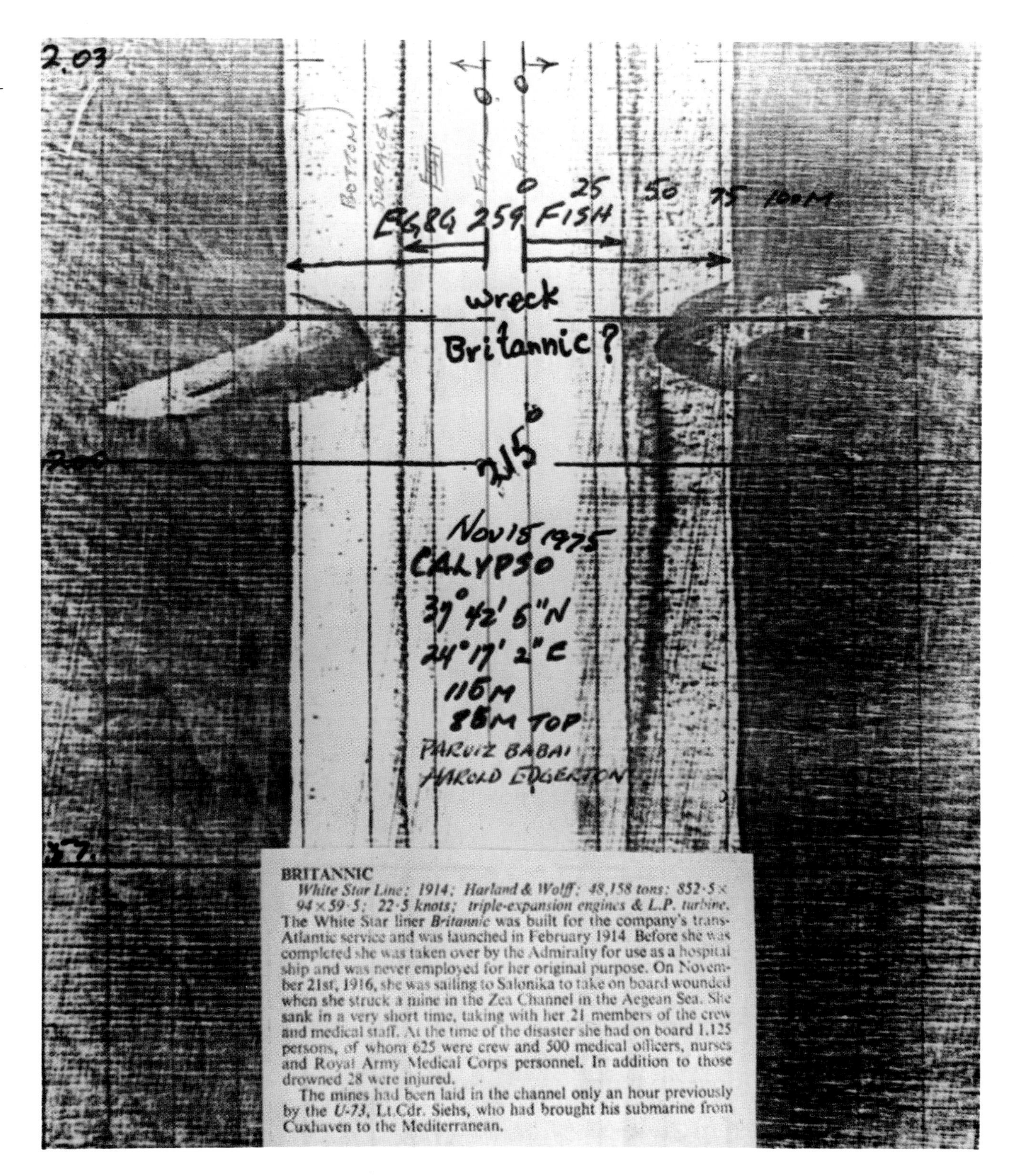

BRITANNIC

White Star Line; 1914; Harland & Wolff; 48,158 tons; 852·5 × 94 × 59·5; 22·5 knots; triple-expansion engines & L.P. turbine.
The White Star liner *Britannic* was built for the company's trans-Atlantic service and was launched in February 1914 Before she was completed she was taken over by the Admiralty for use as a hospital ship and was never employed for her original purpose. On November 21st, 1916, she was sailing to Salonika to take on board wounded when she struck a mine in the Zea Channel in the Aegean Sea. She sank in a very short time, taking with her 21 members of the crew and medical staff. At the time of the disaster she had on board 1,125 persons, of whom 625 were crew and 500 medical officers, nurses and Royal Army Medical Corps personnel. In addition to those drowned 28 were injured.

The mines had been laid in the channel only an hour previously by the *U-73*, Lt.Cdr. Siehs, who had brought his submarine from Cuxhaven to the Mediterranean.

Discovery of the *Britannic*

The sonar record of the *Britannic*, found by the crew of the *Calypso* using Edgerton side-scan sonar.

The Sonar "Fish"

Harold Edgerton affectionately holds a side-scan sonar "fish" on the deck of an Army LCU. The recorder was installed in the trailer in the rear. John Newton (right), then of Duke University Marine Lab at Beaufort, North Carolina, now director of the Monitor Research and Recovery Foundation in Norfolk, Virginia, holds one of the many buoys that were used to pinpoint the *Monitor* wreck. The sonar, which has a 200-meter range to both sides, covers a square mile in one hour when the ship proceeds at 4 knots.

Cousteau, Baibi, and Papa Flash

Jacques Cousteau, Parviz Baibi, and Harold Edgerton on the *Calypso* with side-scan sonar EG&G "fish" that pinpointed the *Britannic* in Grecian waters in 1975. Success is evident in the men's smiles.

Wreck in the Bay of Patras

An exceptionally clear side-scan sonar "picture" of a wreck in the Bay of Patras, Greece. This unidentified ship appears to have a single mast. It may have been a German supply ship that was lost during World War II.

This side-scan sonar record of the wreck was made on a windy day; the lines show the effects of the waves. Note that the aspect is almost 180 degrees from the other side-scan picture. In the photograph there appears to be an illusion of several masts due to the motion of the sonar caused by the waves.

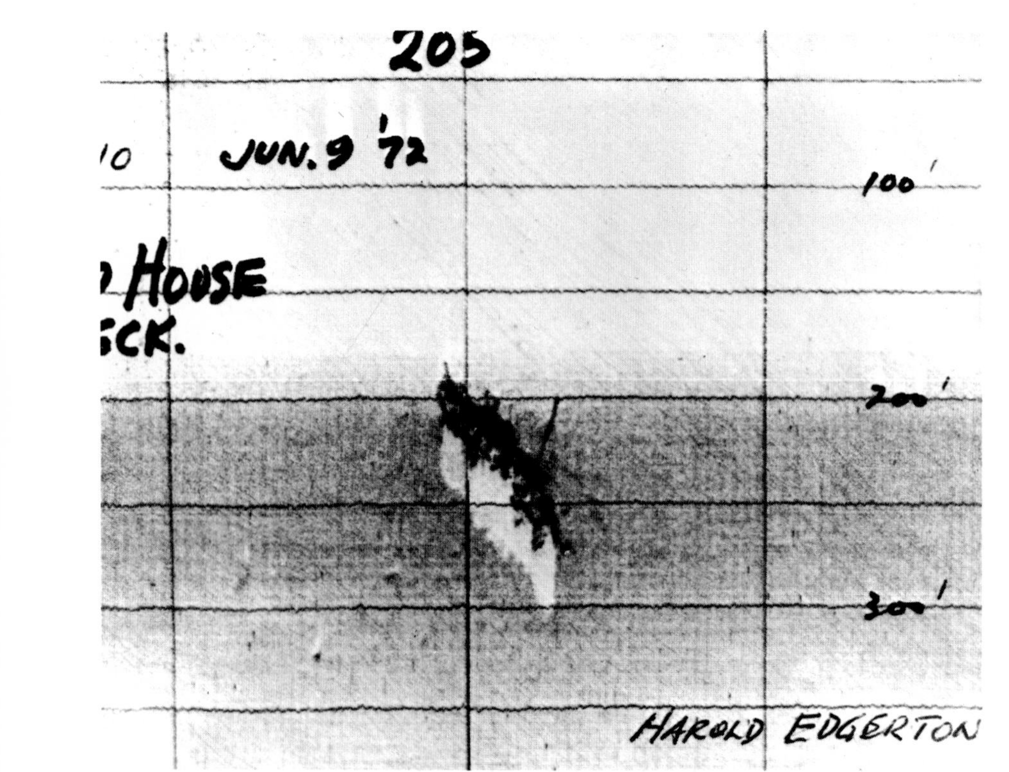

Mysterious Mound

A sonar-revealed mound in the harbor of Sami of the Greek Island of Cephalonia. Could this be an ancient shipwreck? The location of the mound is very close to Ithaca, the home island of Ulysses.

Sonar record made with Dr. Niki Stavrolakes in July, 1973

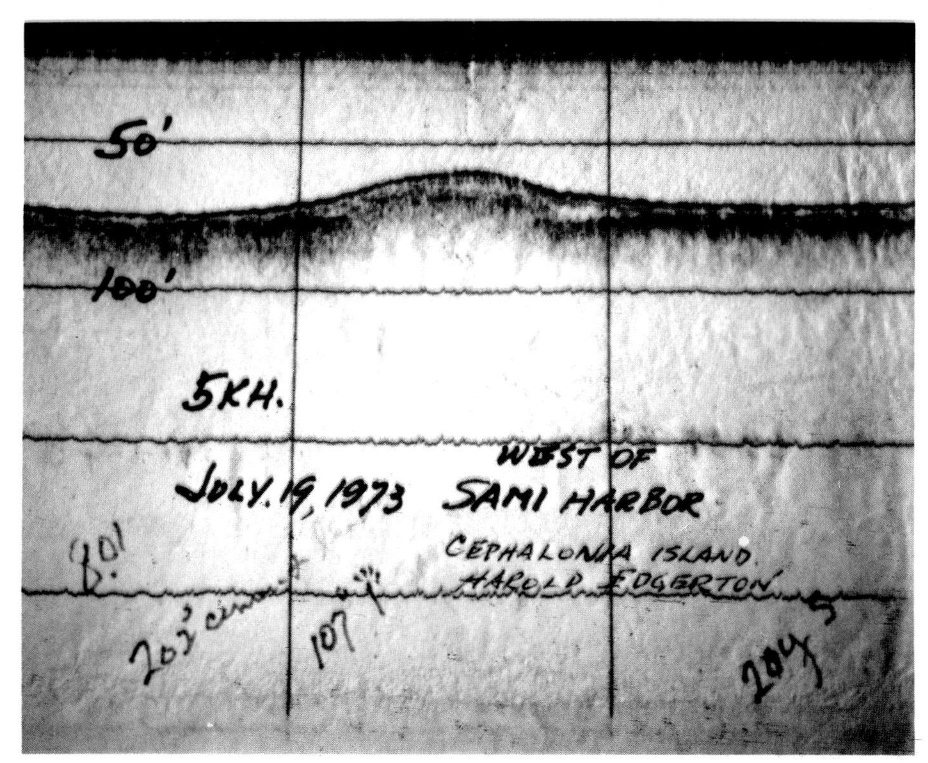

On Looking Below the Bottom of Boston Harbor

Working sonar record made from small ship cruising the harbor area over the Sumner and Callahan Tunnels

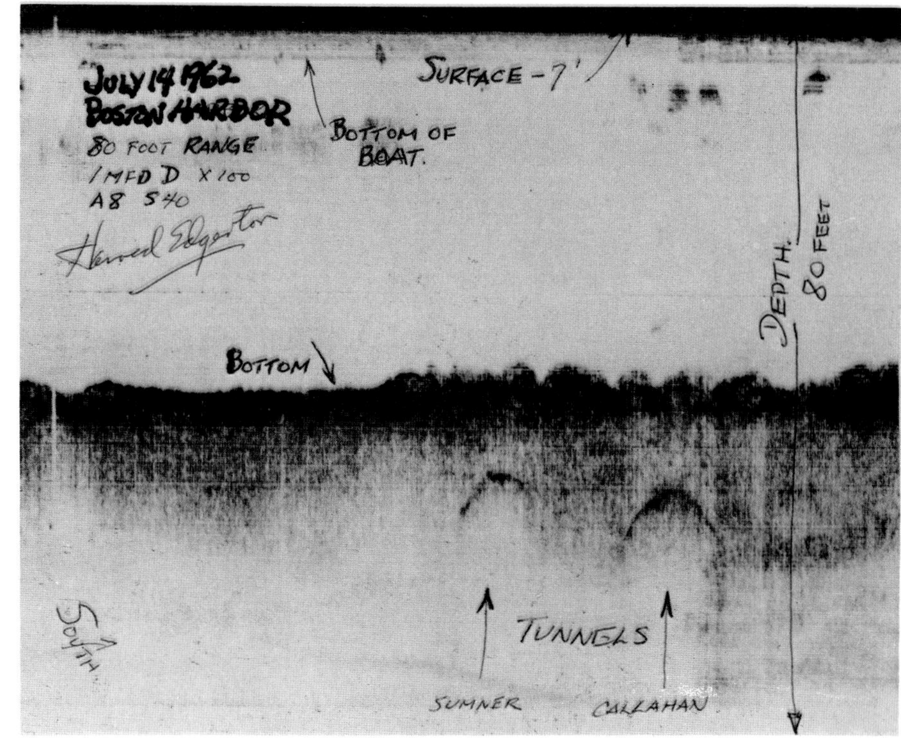

Sonar Record (pinger probe) of Section of Boston Harbor

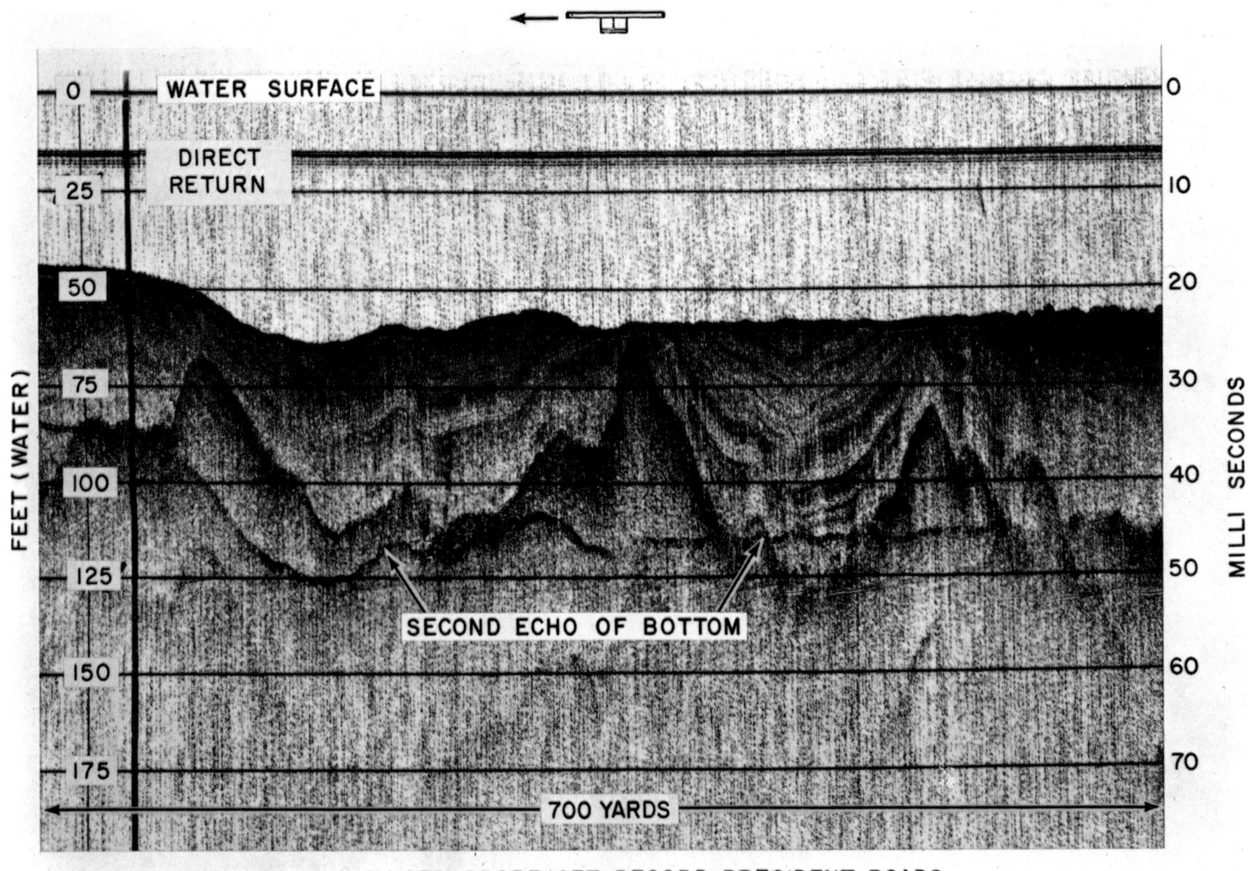

And Something More

Bubble Chamber Arabesques

Physicists at Stanford called on the strobe to catch the evanescent hydrogen bubble trails left by nuclear particles that passed through the bubble chamber.

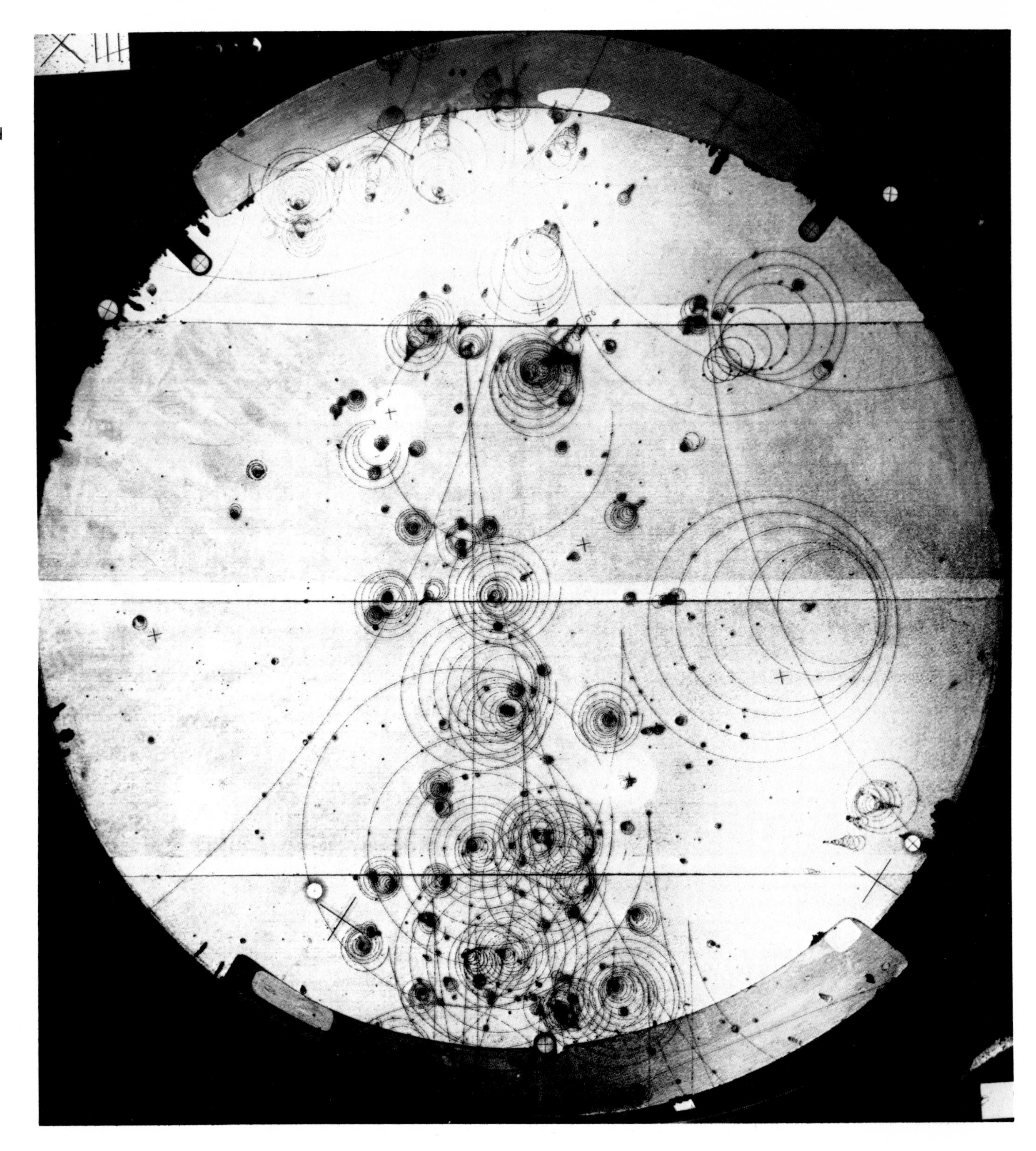

Blood Flow in the Conjunctiva of Man

Electronic flash close-up of the blood circulation of a small portion of the conjunctiva of the eye. This technique provides one of the few external methods of observing blood flow in the microvasculature of man.

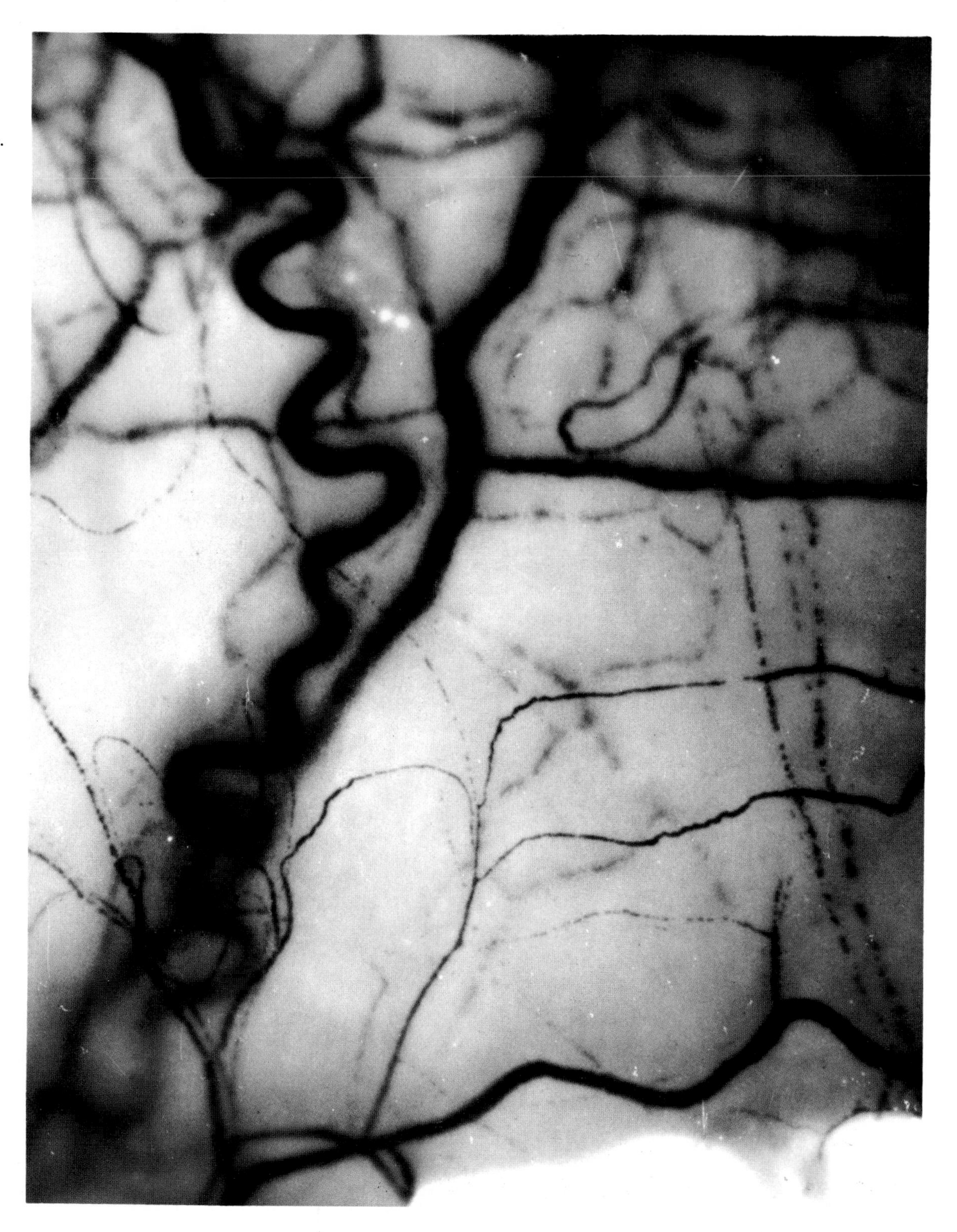

Fusion Experiments Use Xenon Flash Lamps

By far the largest and most ambitious application of flash lamps at this time is being experimented with at the Lawrence Livermore Laboratory in California. Appropriately named SHIVA, after the multiple-armed Hindu god, the equipment, as illustrated by the photograph of the model, consists of 20 laser beam devices that bombard a small pellet simultaneously. The goal is to produce adequate pressure and temperature to induce fusion and generate useful energy.

In technical language, each of the 20 beams, created by light from xenon lamps, produces 10,200 joules of focused laser energy in a 0.095 microsecond pulse averaging 510 joules of energy on the target. A total of 2,080 flash lamps (44-inch arc length, 15-mm diameter, and 2-mm wall thickness) are used in the 20 disc amplifiers and 244 flash lamps of 15-inch length in the other parts of the arrangement.

The total energy stored in the capacitors for the 20 beams is 25 x 10^6 joules. The bank uses 7,600 capacitors of 14.5 microfarads and 800 capacitors of 25 microfarads totaling 0.13 farads at 20 kilovolts.

A proposal for the next step, Shiva Nova, calls for an increase of 20 to 30 times the energy of the previous experiment.

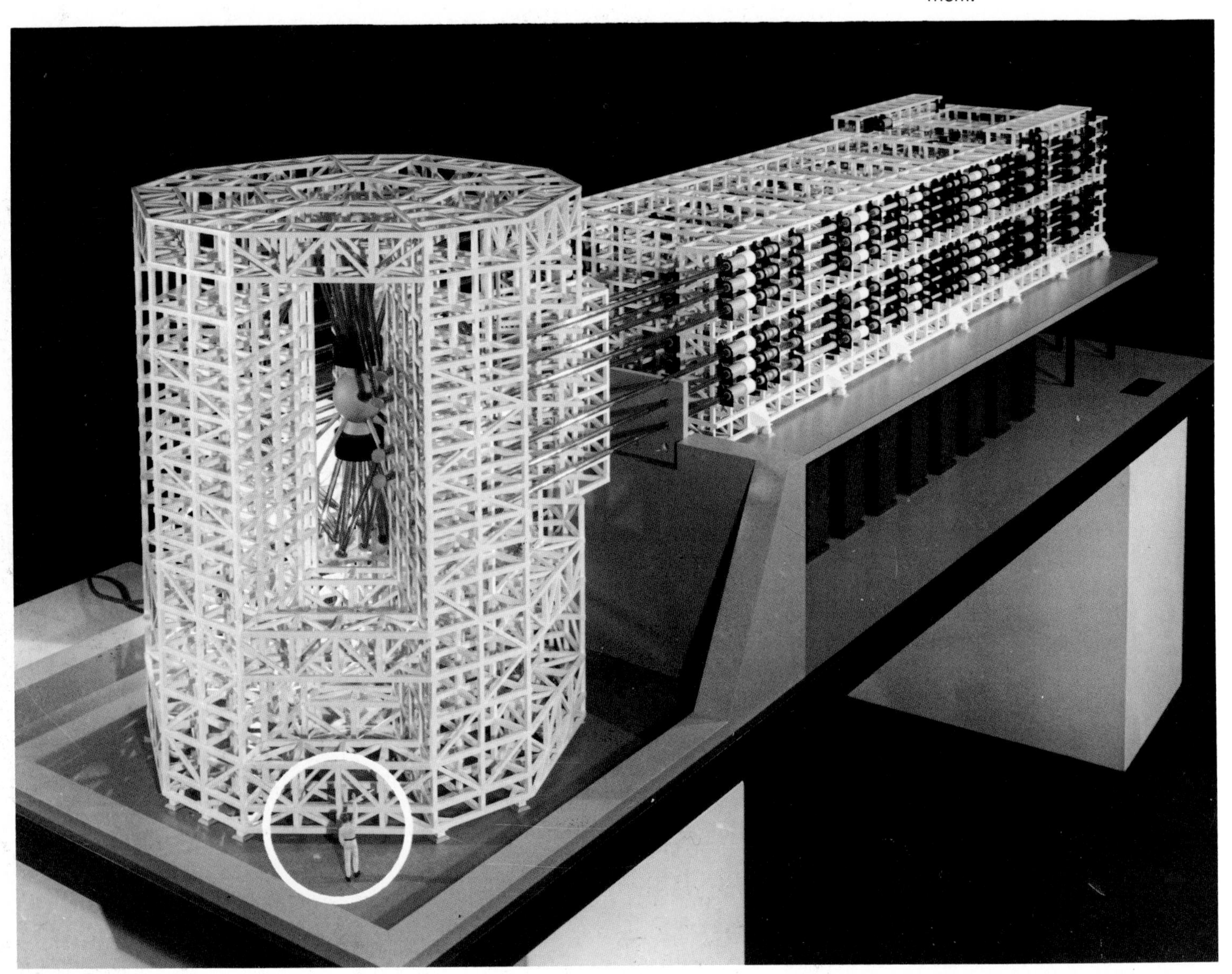

Coda — Setting Sun

Six stages of the setting sun through the grillwork adorning the entrance to MIT's Rogers Building. The pictures were taken five minutes apart. Twice a year at sunset the sun aligns directly with the central corridor of MIT.

Postscript

A reader of the manuscript for this book fairly complained that in no place could he find any photograph of a stroboscope. He was curious to see a picture of Edgerton's magic lamp. On the following pages the authors oblige with several illustrations of strobes and other examples of equipment discussed in this book.

Dramatis Personae

Along with the instruments are the three men who have worked over the years to further the drama of the stroboscopic revolution, Edgerton, Germeshausen, and Grier.

Professor Edgerton acknowledges that he is not entirely happy with these pictures of stroboscopes. Although they appear neatly packaged in the photographs, the real beauties of a stroboscope are found in the sophisticated and intricate technology inside them. Readers who wish to investigate the stroboscope in depth should acquire a copy of *Electronic Flash, Strobe*, by H. E. Edgerton, a technical text published in 1970 by McGraw-Hill Book Company, New York, and reprinted by the MIT Press in the spring of 1979. This technical book has an extensive bibliography about electronic flash lamps, circuits, and applications.

New Lamp

Edgerton associates, Herbert Grier and Fred Barstow, examine a new xenon flash lamp.

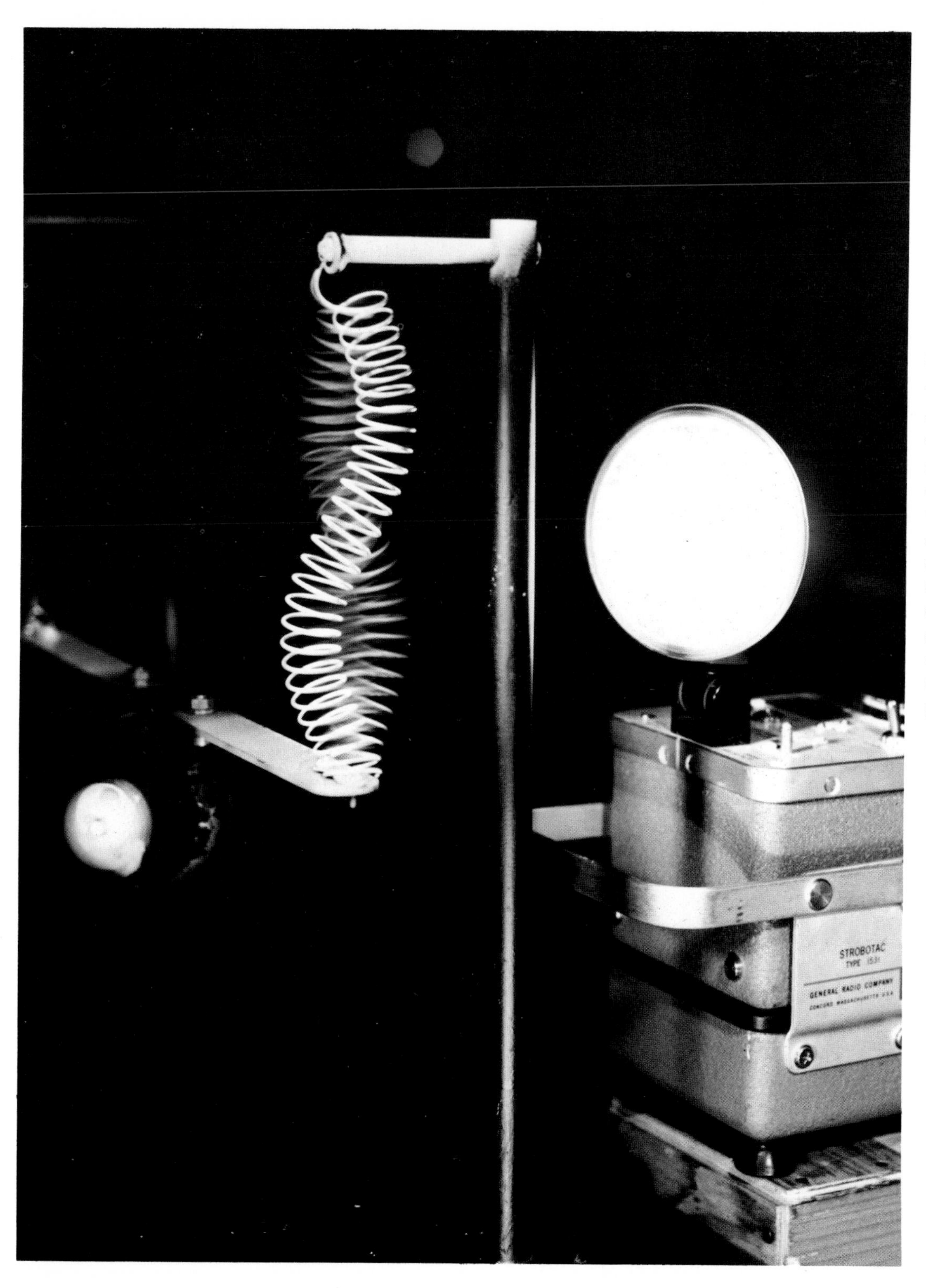

Spring Dance

This stroboscope is shown in the art of photographing the violent oriental dance of a spring. The resonant vibration of the spring is set up by an unbalanced weight on a motor. A stroboscope adjusted to the same frequency shows the shape of the spring. The stroboscope (Strobotac) is calibrated so that the resonant frequency can be measured and the form of the vibration observed and photographed.

Strobe Display

Kenneth J. Germeshausen contemplates a hydraulic strobe display of periodic water drops which apparently makes the drops stop or even climb slowly from the bucket into the nozzle. The GenRad strobe uses a xenon strobotron lamp developed by him. The lamp has numerous applications in science and industry where short duration pulsing radiation is required.

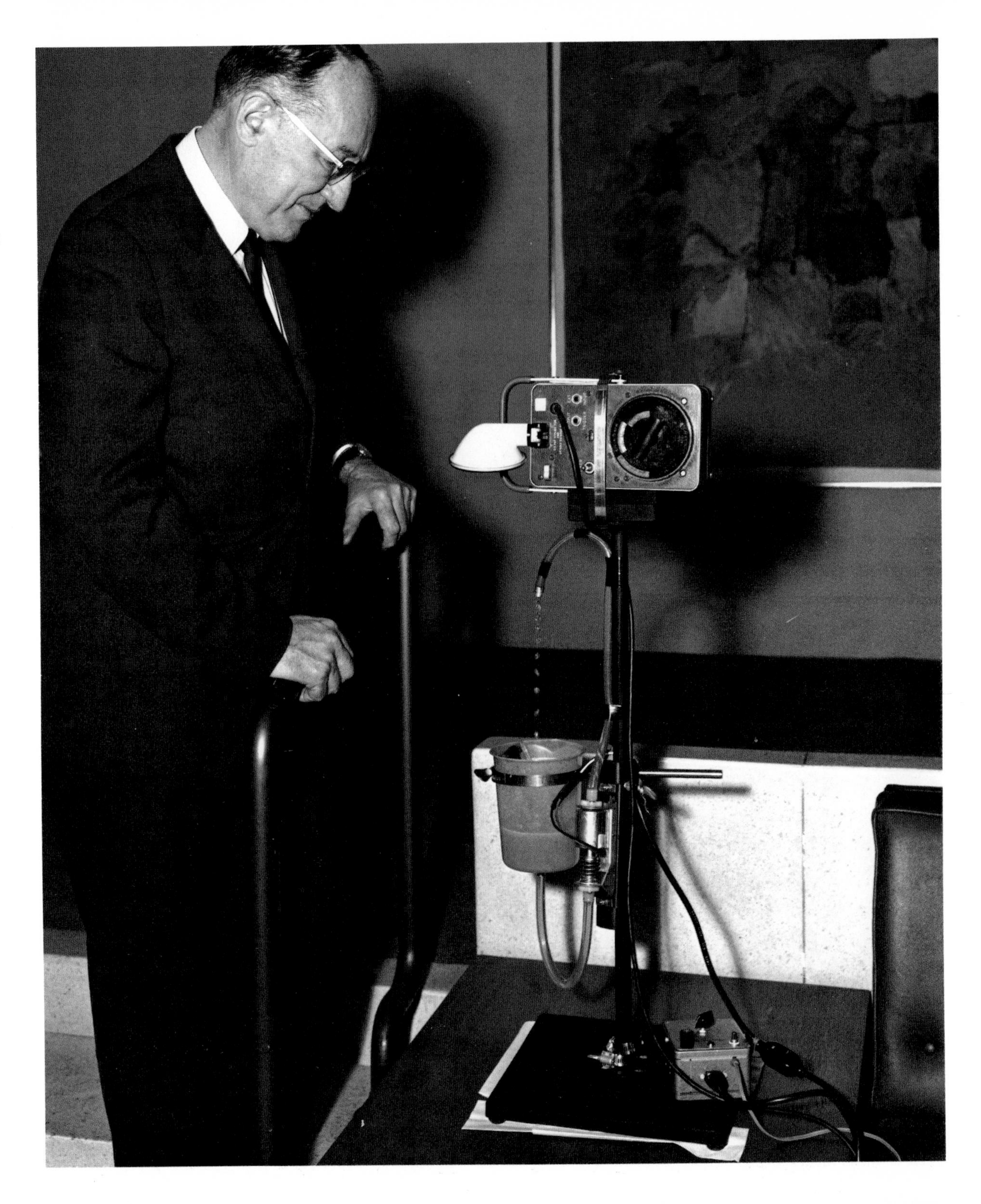

Partners

Harold Edgerton (center) and his partners Herbert Grier (left) and Kenneth Germeshausen (right) discuss the problems of night aerial photography using giant quartz flash lamps such as the one on the table.

Eisenstaedt

Distinguished photo-journalist Alfred Eisenstaedt (left) was entranced by this hydraulic strobe display when Edgerton demonstrated his equipment in the MIT Strobe Lab. The apparent upward flow of the drops, from the bucket back to the nozzle, is due to the regular spacing of the drops and the frequency of the strobe flashes.

Surprised Owl

A great horned owl is released from his perch by Charles Miller and flies toward the white bar while Edgerton directs the camera. Two strobe lights, one at right and another out of sight on the left, provide the multiple flashes. The ordinary light in the center is used for illumination, but also slightly blurs the action. The owl was brought to the multi-flash lab from Boston's Museum of Science by Susan Lockwood (center).

Calvin Campbell

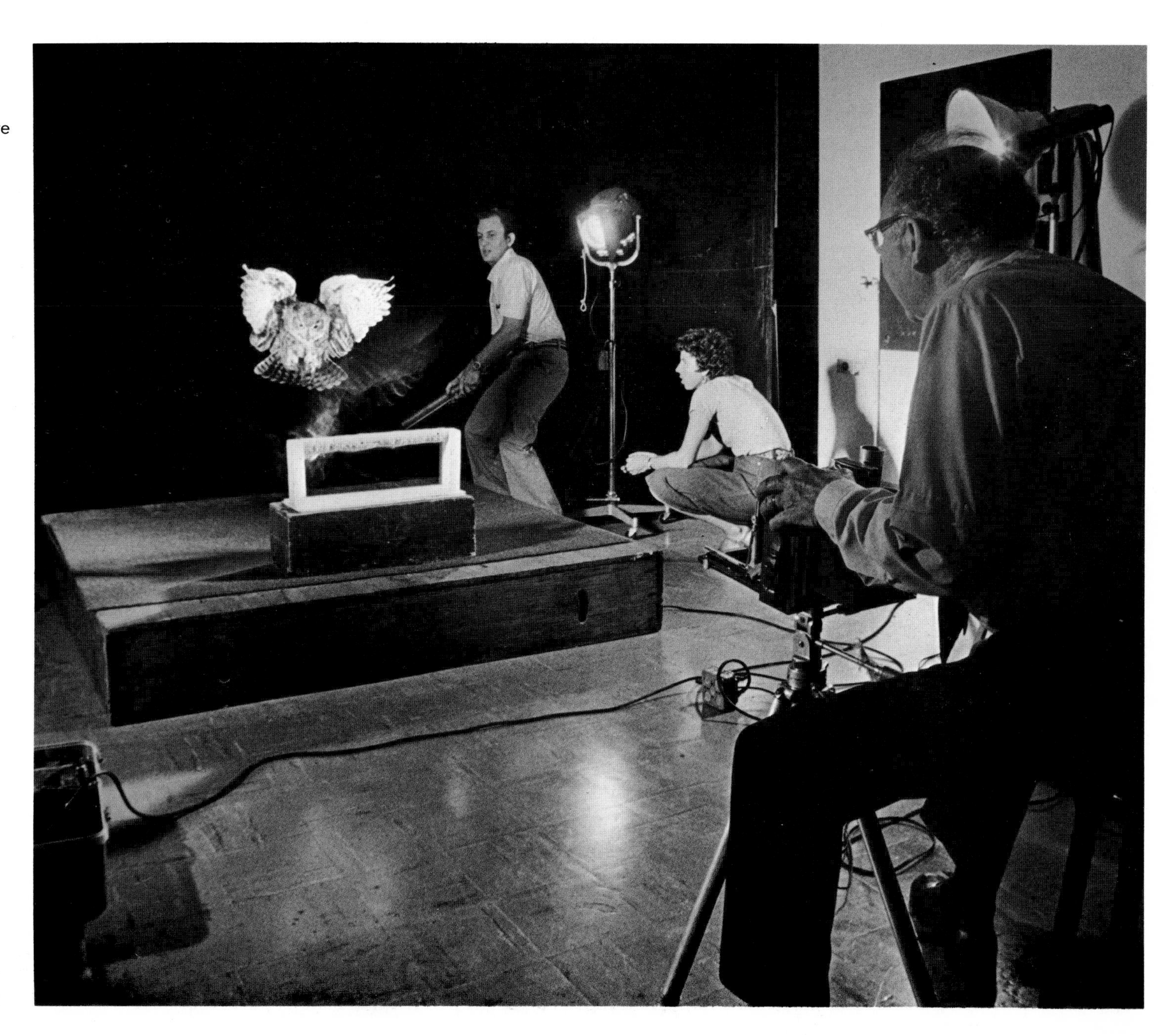

Papa Flash and a Lamp

Edgerton prepares to take underwater photographs of the sea bottom off Jamaica. An experimental electronic flash lamp and complete circuit with a battery are shown in a cylindrical watertight container attached to a Nikonos underwater camera.

Behemoth Flash Lamp

This 40,000 watt-second xenon flash lamp is one of the largest ever built and was designed for use in aerial reconnaissance during World War II. Two of these strobe lamps were used to take the photograph of MIT's Great Dome from across the Charles River. Edgerton and strobe lab technician Vernon E. "Bill" MacRoberts examine the flash tube.

Charles E. Miller

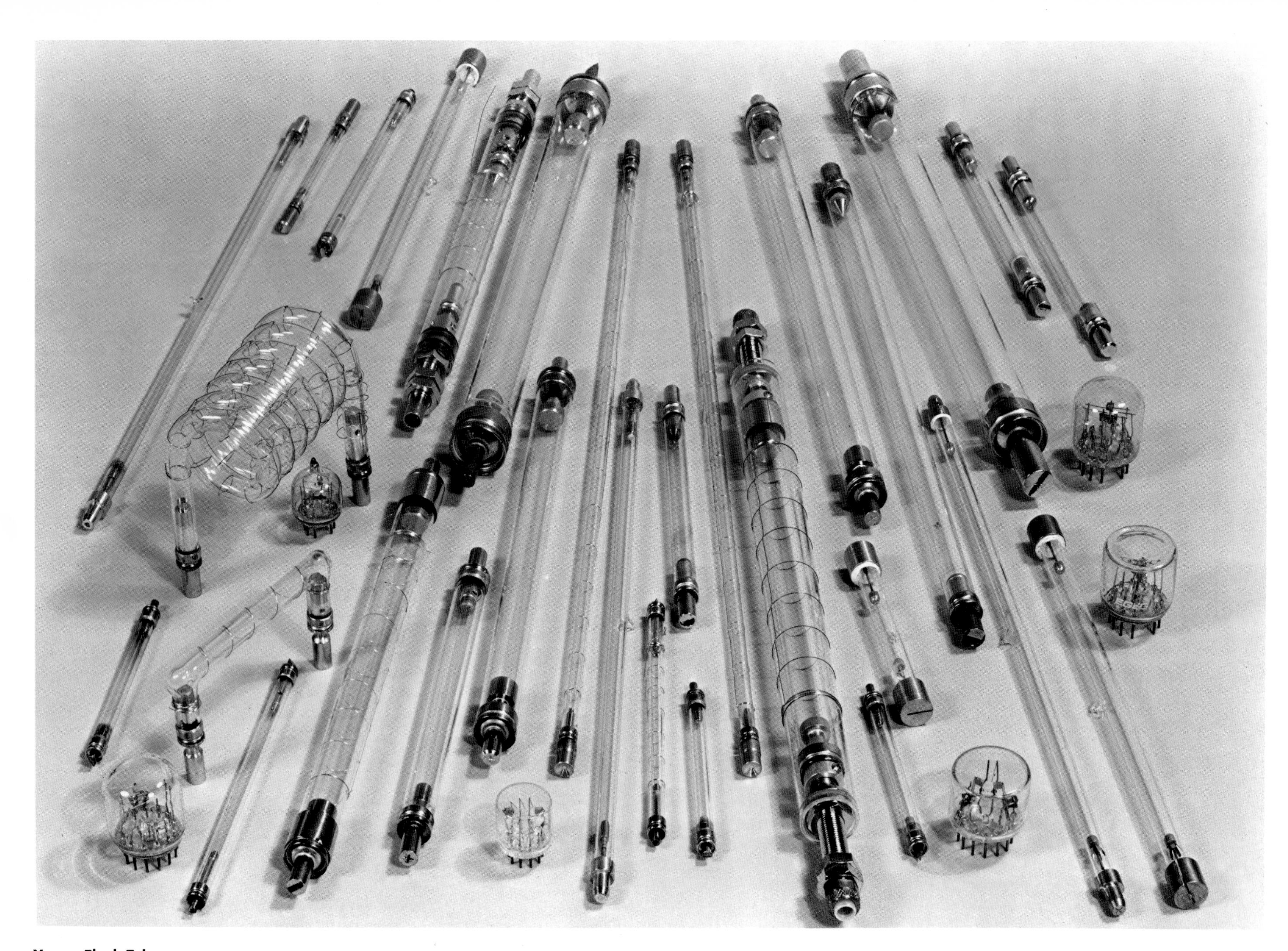

Xenon Flash Tubes

These tubes, of various shapes and sizes, are currently being used in high-speed photography.
EG&G

Notes

1. Kenneth Clark, *The Other Half* (New York: Harper & Row, 1977), p. 189. Berenson was speaking of his old friend, the King of Sweden, and Clark speaks of him as having "a sweetness and simplicity of character which I have rarely seen equalled." The title of this book, *Moments of Vision*, was suggested by the title of a volume of poems by Thomas Hardy mentioned in Clark's book. Clark said that he had used "Moments of Vision" as the title of the best lecture he ever gave.

2. Beaumont Newhall, *The History of Photography* (New York: Museum of Modern Art, 1964) (Distributed by Doubleday & Company, Garden City, New York), p. 173.

3. Antonio Giulio Bragaglia, *Fotodinamismo Futurista*, 3rd ed. (Torino, Italy: Giulio Einaudi, 1970).

4. Gustaf Cavallius, *Velasquez' Las Hilanderas: An Explication of a Picture Regarding Structure and Association* (Uppsala, Sweden: Almqvist & Wiksell, 1972), p. 143.

5. A. M. Worthington, *A Study of Splashes* (London: Longmans, Green, and Co., 1908), p. 1.

6. Ibid., pp. 118-119.

7. Published in *National Geographic*, September 1975, p. 339.

8. C. M. Breder, Jr., *Zoologica* 26, part 4 (December 29, 1941), N.Y. Zoological Society. Contains photographs of flying fish and a technical discussion by the author.

9. C. M. Breder, Jr., *Annals of the New York Academy of Sciences* 43, article 4, pp. 145–172. Analysis of wavelike motions and other actions of the sea horse *(Hippocampus)*.

Reading List

1. Allen, Arthur A. *Stalking Birds With Color Camera.* Washington, D. C.: National Geographic Society, 1951.

2. Bleitz, Don. *Birds of the Americas.* Scheduled for publication in 1979 by Bleitz Wildlife Foundation, 5334 Hollywood Ave., Los Angeles, CA 90027.

3. Dalton, Stephen. *Borne on the Wind: The Extraordinary World of Insects.* New York: Reader's Digest Press, 1975.

4. Edgerton, Harold E., and Killian, James R., Jr. *Flash! Seeing the Unseen by Ultra High-Speed Photography.* Boston, Mass.: Hale, Cushman and Flint, 1939.

5. Edgerton, Harold E., and Killian, James R., Jr. *Flash! Seeing the Unseen by Ultra High-Speed Photography.* 2nd rev. ed. Boston, Mass.: Charles T. Branford Company, 1954.

6. Edgerton, Harold E. *Electronic Flash, Strobe.* New York: McGraw-Hill Book Company, 1970. Reprinted by The MIT Press, 1979. This technical book has an extensive bibliography about electronic flash lamps, circuits, and applications. Some historical information is also given.

Bibliographic information is also given for the magneto-optic shutter, used to photograph the atomic bomb explosions.

7. Edgerton, Harold E. "Silhouette photography of small active subjects." *Journal of Microscopy* 110, part 1 (May 1978): 79–81.

8. Früngel, Frank. *High Speed Pulse Technology*, 3 vols. New York: Academic Press, 1965+. Excellent bibliography.

9. Gernsheim, Helmut, in collaboration with Gernsheim, Alison. *A Concise History of Photography.* New York: Grosset & Dunlap, 1965.

10. Greenewalt, Crawford H. *Hummingbirds.* Garden City, New York: Doubleday & Company, Inc. for the American Museum of Natural History, 1960.

11. Kepes, Gyorgy. *Language of Vision.* Chicago: Paul Theobald & Co., 1944.

12. Kepes, Gyorgy. *The New Landscape in Art & Science.* 3rd printing. Chicago: Paul Theobald & Co., 1963.

13. Laporte, Marcel. "Les Lampes à Eclair Lumiére Blanche et Leur Applications, *Gauthier-Villers.* Paris: 1949. Contains a bibliography of early work from 1936. See the bibliography of chapter 3 in reference no. 4 of this reading list for more of Laporte's publications.

14. Leen, Nina. *The Bat.* New York: Holt, Rinehart & Winston, Inc., 1976.

15. Leen, Nina. *Snakes.* New York: Holt, Rinehart & Winston, Inc., 1978.

16. Marshak, I. S. *Electronic Flash* (in Russian). Moscow, 1963. Translated from the Russian by Scripta Publishing Company, Division of Scripta Technica, Inc., 1511 K Street, N.W., Washington, D.C. 20005. Many references.

17. Newhall, Beaumont. *The History of Photography.* New York: Museum of Modern Art, 1964.

18. *Photography as a Tool.* New York: Time-Life Books, 1971.

19. Shapiro, S. L., ed. *Ultrashort Light Pulses, Picosecond Techniques and Applications.* Berlin, Heidelberg, New York: Springer-Verlag, 1977.

20. Today, the family of flash lamps—mercury, argon, krypton, xenon, etc.—continues to expand. Up-to-date information appears in the catalogues and publications of EG&G, Inc. (35 Congress Street, Salem, MA 01970, USA) as well as other publications. The end is not in sight.

A Note About the Authors

Harold E. Edgerton is Institute Professor Emeritus of Electrical Engineering at MIT.

James R. Killian, Jr., president of MIT from 1949 to 1959, is Honorary Chairman of the MIT Corporation.

About the Authors

Harold Edgerton is Institute Professor Emeritus of Electrical Engineering at MIT and continues to add to his photographic work. James R. Killian, Jr., is Honorary Chairman of the MIT Corporation, President Emeritus of MIT, and author of *Sputnik, Scientists, and Eisenhower* (MIT Press 1977).

Now available from The MIT Press in a paperback edition:

Electronic Flash, Strobe
by Harold E. Edgerton

"*Electronic Flash, Strobe* is a very special volume. It is written by the man who pioneered the method, who made the equipment, who made the early action-stopping photographs that amazed and inspired us, and who has 40 years of personal participation to report. . . . Since one seldom finds so authoritative a writer in any field, it would be wise to pick this particular volume up and give it your thorough attention." —*Modern Photography*

The MIT Press
Massachusetts Institute of Technology
Cambridge, Massachusetts 02142